TurboCAD 2019 For Beginners

Tutorial Books

TurboCAD 2019 For Beginners

© Copyright 2020 by Tutorial Books

This book may not be duplicated in any way without the express written consent of the publisher, except in the form of brief excerpts or quotations for review. The information contained herein is for the personal use of the reader. It may not be incorporated in any commercial programs, other books, databases, or any software without the written consent of the publisher. Making copies of this book or any portion for a purpose other than your own is a violation of copyright laws.

Limit of Liability/Disclaimer of Warranty:

The author and publisher make no representations or warranties concerning the accuracy or completeness of the contents of this work and expressly disclaim all warranties, including without limitation warranties of fitness for a particular purpose. The advice and strategies contained herein may not be suitable for every situation. Neither the publisher nor the author shall be liable for damages arising there onwards.

Trademarks:

All brand names and product names used in this book are trademarks, registered trademarks, or trade names of their respective holders. The author and publisher are not associated with any product or vendor mentioned in this book.

TurboCAD 2019 For Beginners

For resource files and technical support, contact us at:
online.books999@gmail.com

TurboCAD 2019 For Beginners

TurboCAD 2019 For Beginners

Table of Contents

Introduction x
    Scope of this Book x

Chapter 1: Introduction to TurboCAD 2019 1
    Introduction 1
    System requirements 1
    Starting TurboCAD 2019 1
        Starting a new drawing 2
    TurboCAD user interface 3
        File Menu 3
        Graphics Window 4
        Menu Bar 4
        Toolbar 4
        Inspector Bar 6
        Status bar 6
        Dialogs and Palettes 7
        Local Menus 7
    Selection Modes 9
    Saving a drawing 11
    Help 12
    Keyboard Shortcuts 12

Chapter 2: Drawing Basics 13
    Drawing Basics 13
        Drawing Lines 13
        Erasing, Undoing and Redoing 15
        Drawing Rectangles 16
        Drawing Polygons 18
        Drawing Polylines 20
        Drawing Circles 21
        Drawing Arcs 23
        Drawing Splines 28
        Drawing Ellipses 30
    Exercises 32

Chapter 3: Drawing Aids 35
    Drawing Aids 35
        Setting Grid and Snap 35
        Specifying the Angle for Ortho mode 36
        Using Layers 37
        Using Local Snaps 39
        Running Snap Modes 41

TurboCAD 2019 For Beginners

    Using Auxiliary Lines 42
    Using Zoom tools 42
    Panning Drawings 43
  Exercises 44

Chapter 4: Editing Tools 45
  Editing Tools 45
    The Move tool 45
    The Vector Copy tool 46
    The Rotate tool 47
    The Scale tool 48
    The Trim tool 49
    The Shrink/Extend tool 50
    The Fillet 2D tool 51
    The Chamfer tool 52
    The Mirror Copy tool 52
    The Explode tool 54
    The Stretch tool 54
    The Radial tool 55
    The Offset tool 57
    The Array tool 58
    Editing Using Grips 60
  Exercises 64

Chapter 5: Multi View Drawings 73
  Multi-view Drawings 73
    Creating Orthographic Views 73
    Creating Auxiliary Views 81
    Creating Named views 91
  Exercises 92
    Exercise 1 92
    Exercise 2 93
    Exercise 3 93
    Exercise 4 93

Chapter 6: Dimensions and Annotations 95
  Dimensioning 95
    Creating Dimensions 95
    Creating a Dimension Style 106
    Adding Dimensional Tolerances 110
    Geometric Dimensioning and Tolerancing 112
    Editing Dimensions by Stretching 115
    Modifying Dimensions using the Properties palette 115
    Matching Properties of Dimensions or Objects 116

- Exercises 117
  - Exercise 1 117
  - Exercise 2 117
  - Exercise 3 118
  - Exercise 4 118
  - Exercise 5 119

Chapter 7: Section Views 121

- **Section Views** 121
  - The Pick Point Hatching tool 121
  - The Hatch tool 129
- **Exercises** 130
  - Exercise 1 130
  - Exercise 2 131

Chapter 8: Blocks, Attributes, and External References 133

- **Introduction** 133
  - Creating Blocks 133
  - Inserting Blocks 133
  - Replacing the Blocks 134
  - Exploding Blocks 135
  - Using the Library 136
  - Editing Blocks 138
  - Defining Block Attributes 138
  - Inserting Attributed Blocks 140
  - Working with External references 141
  - Clipping External References 144
- **Exercise** 145

Chapter 9: Layouts & Printing 147

- **Drawing Layouts** 147
  - Working with Layouts 147
  - Creating Viewports in the Layout 148
  - Creating a Viewport in the ISO A4 Layout 148
  - Creating the Title Block on the Layout 150
  - Plotting/Printing the drawing 152
- **Exercises** 153
  - Exercise 1 153
  - Exercise 2 153

Index 155

TurboCAD 2019 For Beginners

Introduction

CAD is an abbreviation for Computer-Aided Design. It is the process used to design and draft components on your computer. This process includes creating designs and drawings of the product or system. TurboCAD can be used to create two-dimensional (2D) and three-dimensional (3D) models of products. TurboCAD is one of the first CAD software packages. It was introduced in the year 1986. As a student, learning TurboCAD provides you with a significant advantage.

Scope of this Book

The *TurboCAD 2019 for Beginners* book provides a learn-by-doing approach for users to learn TurboCAD. It is written for students and engineers who are interested to learn TurboCAD 2019 for creating designs and drawing of components or anyone who communicates through technical drawings as part of their work. The topics covered in this book are as follows:

- Chapter 1, "Introduction to TurboCAD 2019", gives an introduction to TurboCAD. The user interface and terminology are discussed in this chapter.

- Chapter 2, "Drawing Basics," explores the essential drawing tools in TurboCAD. You will create simple drawings using the drawing tools.

- Chapter 3, "Drawing Aids," explores the drawing settings that will assist you in creating drawings.

- Chapter 4, "Editing Tools," covers the tools required to modify drawing objects or create new objects using the existing ones.

- Chapter 5, "Multi View Drawings," teaches you to create multi-view drawings standard projection techniques.

- Chapter 6, "Dimensions and Annotations," teaches you to apply dimensions and annotations to a drawing.

- Chapter 7, "Section Views," teaches you to create section views of a component. A section view is the inside view of a component when it is sliced.

- Chapter 9, "Blocks, Attributes, and External References," teaches you to create Blocks, Attributes, and Xrefs. Blocks are a group of objects in a drawing that can be reused. Attributes are notes or values related to an object. Xrefs are drawing files attached to another drawing.

TurboCAD 2019 For Beginners

- Chapter 10, "Layouts and Printing," teaches you to create Layouts and annotative objects. Layouts are the digital counterparts of physical drawing sheets. Annotative objects are dimensions, and notes, which their sizes concerning drawing scale.

Chapter 1: Introduction to TurboCAD 2019

In this chapter, you will learn about:

- **TurboCAD user interface**
- **Customizing user interface**

Introduction

TurboCAD is a legendary software in the world of Computer Aided Designing (CAD). It has completed 33 years by 2019. If you are a new user of this software, then the time you spend on learning this software will be a wise investment. I welcome you to learn TurboCAD using this book through step-by-step examples to learn various commands and techniques.

System requirements

The following are system requirements for running TurboCAD smoothly on your system.

- Microsoft Windows 8, Windows 7, Windows 10.
- 8 GB of RAM.

Starting TurboCAD 2019

To start **TurboCAD 2019**, double-click the **TurboCAD 2019** icon on your Desktop. When you double-click the **TurboCAD 2019** icon on the desktop, the **TurboCAD 2019 Workspace Style** dialog appears.

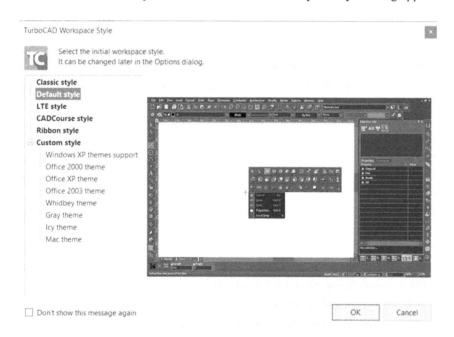

On the **TurboCAD Workspace Style** dialog, select the **Default style**, and then click **OK**.

Starting a new drawing

To start a new drawing, click the **New** button on any one of the following:

- **Main** toolbar
- **File Menu**

The **New TurboCAD 2019 Drawing** dialog appears when you click the **New** button. In this dialog, select the **New from Scratch** for creating a 2D drawing from scratch.

Select the **New from Template** option to create a drawing file using a template.

TurboCAD 2019 For Beginners

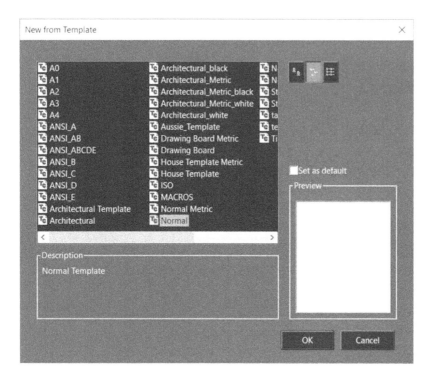

TurboCAD user interface

The drawing file consists of a graphics window, menu bar, toolbars, Inspector Bar, and other screen components, depending on the workspace that you have selected.

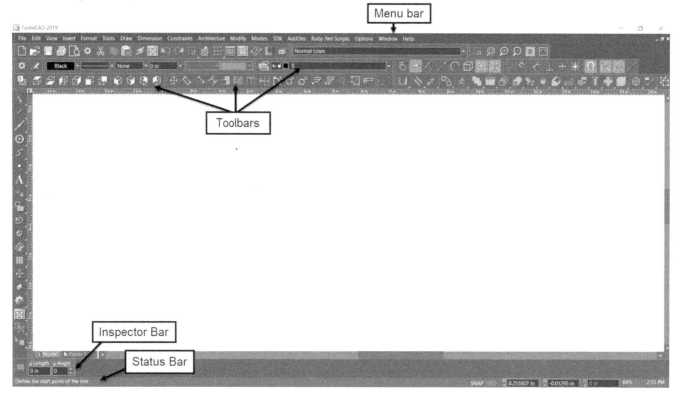

File Menu

Introduction to TurboCAD 2019

The **File Menu** appears when you click on the **File** option located at the top left corner of the window. The **File Menu** consists of a list of options such as **New**, **Open**, **Save**, and **Print** on the left side.

Graphics Window

The Graphics window is the blank space located below the file tabs. You can draw objects and create 3D graphics in the graphics window.

Menu Bar

The Menu Bar is located at the top of the window just below the title bar. It contains various menus such as File, Edit, View, Insert, Format, Tools, Draw, Dimensions, Modify, and so on. Clicking on any of the words on the Menu Bar displays a menu. The menu contains various tools and options. There are also sub-options available on the list. These sub-options are displayed if you click on an option with an arrow.

Toolbar

A toolbar is a set of commands, which help you to perform various operations. Various toolbars available in TurboCAD are given next.

	This toolbar has many options such as New, Open, Save, and Print.

 Draw	This toolbar has commands to create the drawing entities.
 Dimension	This toolbar has commands to apply dimensions to the drawing.
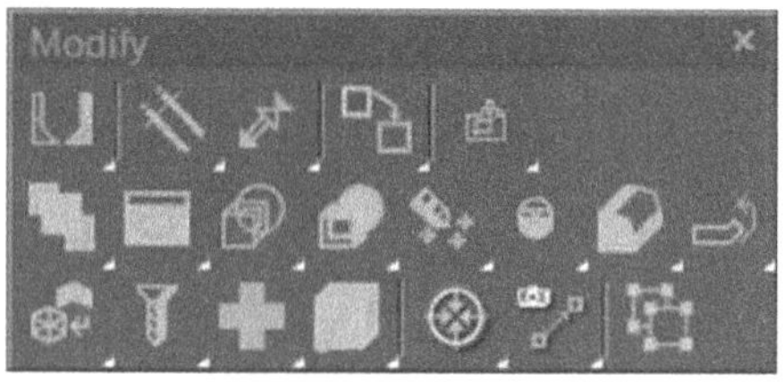 Modify	This toolbar has commands to perform various operations on drawing entities.
 Zoom	This toolbar has options to zoom in or zoom out of the drawing.
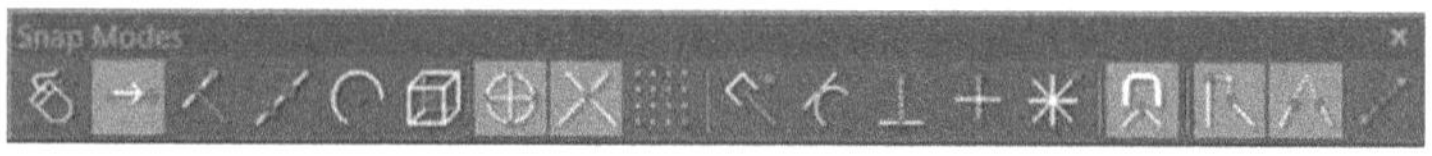 Snap Modes	This toolbar has options that help you to snap to the points in the graphics window. As a result, you can create the drawing entities easily.
 Layers	This toolbar has options to create and select layers.
 Work	This toolbar has commands to customize the workspace.
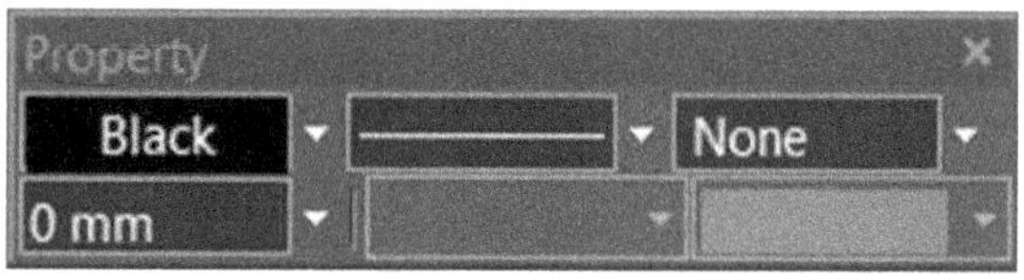 Property	This toolbar has options to specify the color, width, and pattern of the drawing entities.

	This toolbar has options to create named views and change the view orientation.

Some toolbars are not visible by default. To display a particular toolbar, click **View > Toolbars** on the menu bar. Next, select the checkboxes next to the toolbars to be displayed. Click **Close** on the **Customize** dialog.

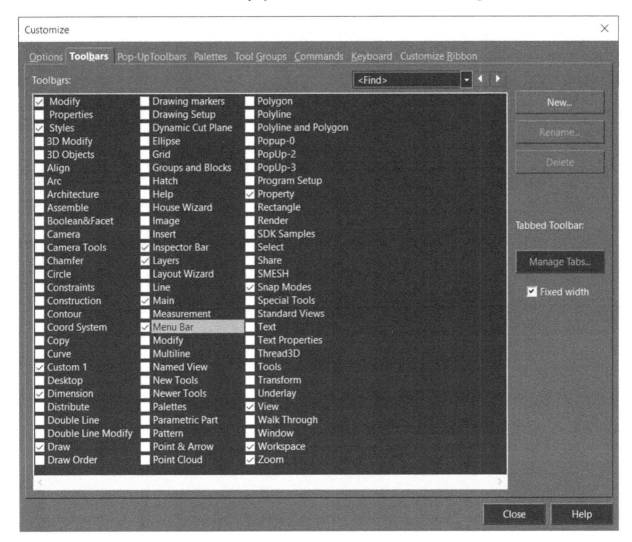

Inspector Bar
It allows you to create objects by entering exact numerical parameters.

Status bar
The status bar is available below the graphics window. It shows the prompts and the action taken while using the commands.

Dialogs and Palettes

Dialogs and Palettes are part of the TurboCAD user interface. Using a dialog or a palette, you can easily specify many settings and options at a time. Examples of dialogs and palettes are as shown below.

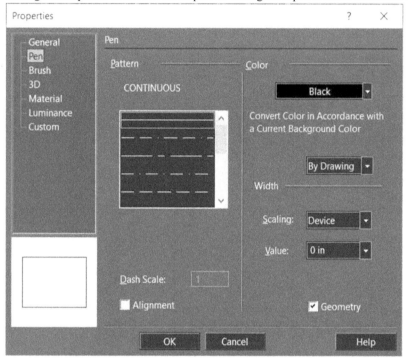

Dialog

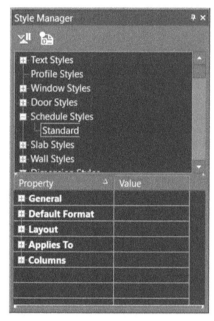

Palette

Local Menus

Local Menus appear when you right-click in the graphics window. TurboCAD provides various Local menus to help you access tools and options very easily and quickly. There are multiple types of of Local menus available in TurboCAD. Some of them are discussed next.

Right-click Menu

This local menu appears whenever you right-click in the graphics window without activating any command or selecting an object.

Select and Right-click menu

This local menu appears when you select an object from the graphics window and right-click. It consists of editing and selection options.

TurboCAD 2019 For Beginners

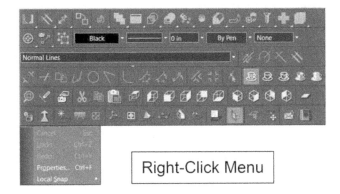

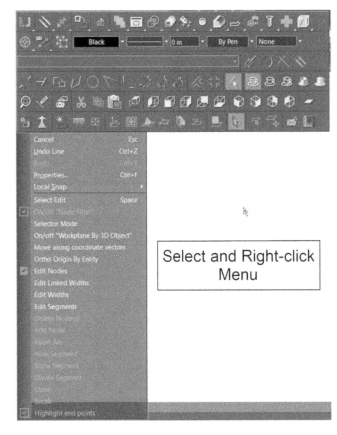

Tool local menu

This local menu appears when you activate a command and right-click. It shows options depending upon the active command. The local menu below shows the options related to the Rectangle tool.

TurboCAD 2019 For Beginners

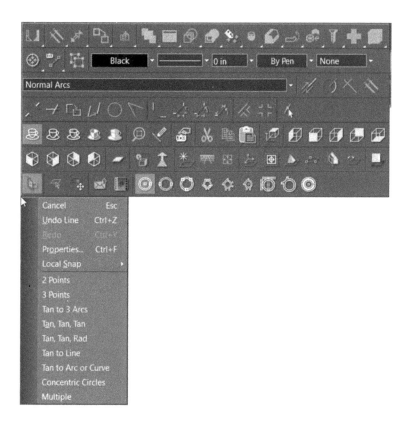

Selection Modes

Use the **Select** tool to select the entities of a drawing. To do this, click **Tools > Select** on the Menu bar. Next, click on the entities to select. You can select multiple elements of a drawing using a selection mode. You can select various elements by using four types of selection modes. The first type is a **Rectangular Mode**. You can create this type of selection window by defining its two diagonal corners. When you set the first corner of the selection window on the left and second corner on the right side, the elements which fall entirely under the selection window will be selected.

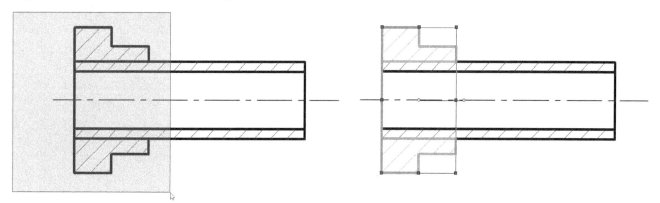

However, if you define the first corner on the right side and the second corner on the left side, the elements, which fall entirely or partially under the selection window, will be selected.

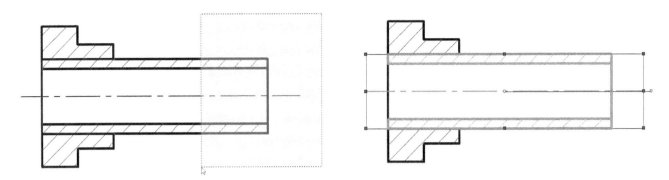

The second type of selection mode is **Window Polygon Mode**. Activate the **Select** tool and click the **Window Polygon Mode** icon on the Inspector Bar. Next, click to specify the first point of the window polygon. Next, press and hold the left mouse button and drag the pointer. Click to specify the second point. Next, move the pointer across the elements to be selected and keep clicking. Double-click to close the window polygon.

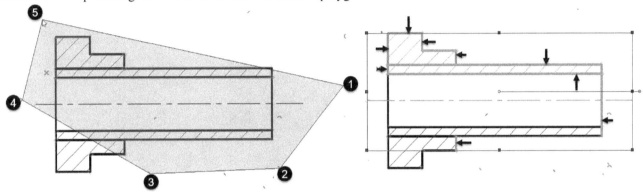

The third type of selection mode is **Crossing Polygon Mode**. If you create a crossing polygon, the elements which fall wholly or partially under the polygon will be selected.

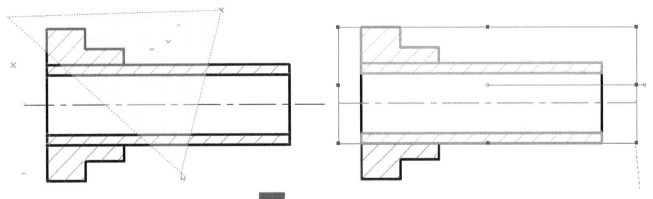

The fourth type of selection mode is **Fence Mode**. Select this option and create a fence crossing the elements of the drawing.

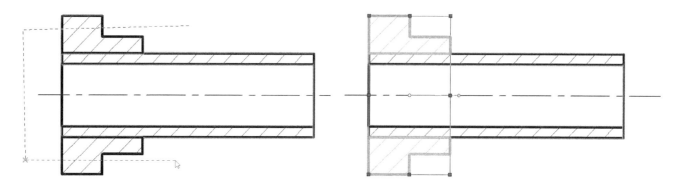

Saving a drawing

You can save a drawing by clicking the **Save** icon on the **Main** toolbar (or) click **File > Save** on the menu bar. On the **Save As** dialog, browse to the desired location and enter the name of the file. Next, select the Save as type, and then click **Save**.

TurboCAD 2019 For Beginners

Help

Press F1 on your keyboard; the TurboCAD 2019 User Guide appears. On the TurboCAD 2019 User Guide window, enter a keyword in the **Search** bar located at the top-right corner; the related topic appears.

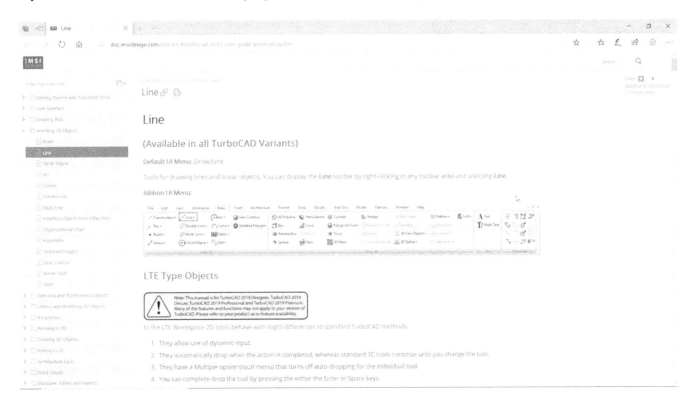

Keyboard Shortcuts

You can view the keyboard shortcuts by clicking **Help** > **Keyboard** on the Menu bar.

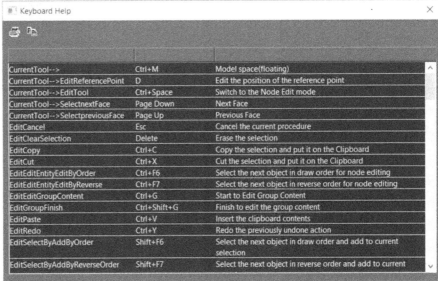

12 | Introduction to TurboCAD 2019

Chapter 2: Drawing Basics

In this chapter, you will learn to do the following:

- **Draw lines, rectangles, circles, ellipses, arcs, polygons, and polylines**
- **Use the Erase, Undo and Redo tools**

Drawing Basics

This chapter teaches you to create simple drawings. You will create these drawings using the essential drawing tools. These tools include **Line**, **Circle**, **Polyline**, and **Rectangle**, and they are available in the **Draw** toolbar, as shown below.

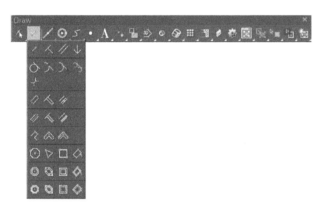

Drawing Lines

You can draw a line by specifying its start point and endpoint using the **Line** tool.

- Start TurboCAD 2019 by clicking the **TurboCAD 2019** icon on your desktop.
- On the **TurboCAD Workspace Style** dialog, select the **Default style** from the left pane and click **OK**. It opens TurboCAD with the Default style interface.

- Select **New from Template** from the **New TurboCAD 2019 Drawing** dialog; the **New from Template** dialog appears. On this dialog, select the **ISO** template and click **OK**.

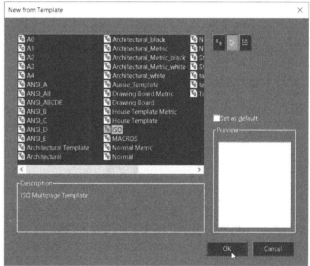

If you click the **New from Scratch** icon on the **New TurboCAD 2019 Drawing** dialog to create a new drawing with the default settings, the default template is the **Normal** template.

- On the Menu bar, click **View > Toolbars**; the **Customize** dialog appears.
- Check the **Snap Modes** and **Zoom** options located at the bottom of the **Toolbars** list.

TurboCAD 2019 For Beginners

- Click **Close** to close the **Customize** dialog.
- Click and drag the **Snap Modes** and **Zoom** toolbars and position them at the location, as shown.

- Click the **Zoom Extents** icon on the **Zoom** toolbar; the entire area in the graphics window will be displayed.
- Deactivate the **Ortho** icon on the **Snap Modes** toolbar.

- To draw a line, click the **Line** icon on the **Draw** toolbar.
- Click in the graphics window to specify the start point of the line.
- Move the pointer toward the right.

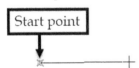

- Press the TAB key.
- Type 50 and press the TAB key.
- Type 0 as the angle value and press ENTER; a new horizontal line of 50 mm length is created.

- Move the pointer upward.
- Press the TAB key.
- Type 20 and press the TAB key.
- Type 90 as the angle value and press ENTER.
- Move the pointer toward the right.
- Press the TAB key.
- Type 100 and press the TAB key
- Type 0 as the angle value and press ENTER.
- Rotate the scroll wheel on your mouse in the forward direction; the drawing is zoomed in.
- Move the pointer downward.
- Press the TAB key.
- Type 20 and press the TAB key.
- Type 270 as the angle value and press ENTER.
- Move the pointer toward the right.
- Press the TAB key.
- Type 50 and press the TAB key.
- Type 0 as the angle value and press ENTER.
- Move the pointer upward.
- Press the TAB key.
- Type 120 and press the TAB key.
- Type 90 as the angle value and press ENTER.
- Move the pointer toward left.
- Press the TAB key.
- Type 50 and press the TAB key.
- Type 180 as the angle value and press ENTER.
- Move the pointer downward.
- Press the TAB key.
- Type 20 and press the TAB key.
- Type 270 as the angle value and press ENTER.
- Move the pointer toward left.
- Press the TAB key.
- Type 100 and press the TAB key.
- Type 180 as the angle value and press ENTER.
- Move the pointer upward.
- Press the TAB key.
- Type 20 and press the TAB key.
- Type 90 as the angle value and press ENTER.
- Move the pointer toward left.
- Press the TAB key.
- Type 50 and press the TAB key.
- Type 180 as the angle value and press ENTER.
- Move the pointer near to the start point of the sketch.
- Right click and select **Close**.

TurboCAD 2019 For Beginners

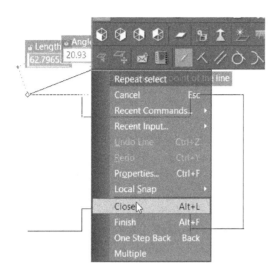

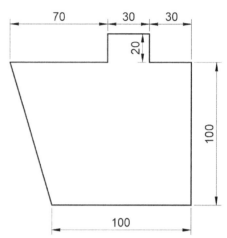

- Click the **Erase** icon on the **Workspace** toolbar.

- Select the lines shown below and press ENTER. It erases the lines.

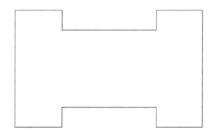

- Click **Save** icon on the **Main** toolbar (or) click **File > Save** on the Menu bar.

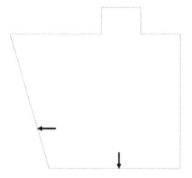

- Browse to a location on your computer.
- Type **Drawing_lines** in the **File name** box.
- Click **Save**.
- Leave the default settings and click **OK** on the **Summary Info** dialog.
- Click **File > Close** on the menu bar to close the file.

- Click the **Undo** button on the **Main** Toolbar. This action restores the lines.

Erasing, Undoing and Redoing
- Click the **New** icon on the **Main** toolbar.
- Select **New from Template** from the **New TurboCAD 2019 Drawing** dialog; the **New from Template** dialog appears. On this dialog, select the **ISO** template and click **OK**.
- Draw the sketch shown below using the **Line** tool. Do not dimension the drawing.

- Click the **Redo** button on the **Main** Toolbar. This action erases the lines again.

Drawing Basics

- Click the **Erase** 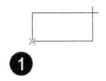 icon on the **Workspace** toolbar.
- Click on the top right corner of the drawing.
- Move the pointer and click at the location, as shown. The lines crossing the selection box are selected.
- Press Enter to delete the selected lines.

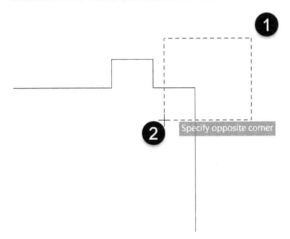

- Pick an arbitrary point in the graphics window.

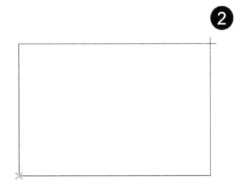

- Move the pointer diagonally toward the right and click to create a rectangle.

Drawing Rectangles

A rectangle is a four-sided single object. You can create a rectangle by just specifying its two diagonal corners. However, there are various methods to create a rectangle. These methods are explained in the following examples.

Example 1
In this example, you will create a rectangle by specifying its corner points.

- Open a drawing new file.
- Click **Line > Rectangle** on the **Draw** toolbar.

Example 2
In this example, you will create a rectangle by specifying its length and width.

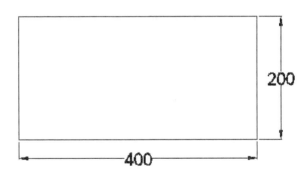

- Click **Line > Rectangle** on the **Draw** toolbar.
- Specify the first corner of the rectangle by picking an arbitrary point in the graphics window.
- Press the TAB Key
- Type **400** in the **Size A** box to specify the length of the rectangle.
- Press the TAB key and type **200** in the **Size B** box to specify the width of the rectangle.

- Press ENTER to create the rectangle.

Example 3

In this example, you will create a rectangle with rounded corners.

- Click **Line** drop-down > **Rectangle** on the **Draw** toolbar.
- Right-click and select the **Fillet corners** option.

By selecting the **Fillet corners** option, the corners of the rectangle will be rounded by filleting. You need to specify the radius of the fillet.

- Specify an arbitrary point in the graphics window.
- Press the TAB key.
- Type **100** in the **Size A** box.
- Press the TAB key and type **80** in the **Size B** box.
- Press the TAB key and type **10** in the **Radius** box.

- Click to create a rectangle with rounded corners.
- Press Esc to deactivate the tool.

Example 4

In this example, you will create an inclined rectangle.

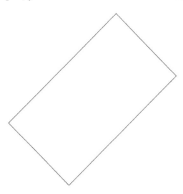

- Click **Line** drop-down > **Rotated Rectangle** on the **Draw** toolbar.

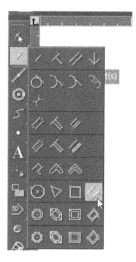

- Specify the first corner of the rectangle by picking an arbitrary point.
- Move the pointer and click to define the length and angle of the rectangle.

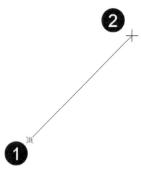

- Again, move the pointer and click to define the width of the rectangle; the rotated rectangle is created.

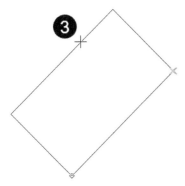 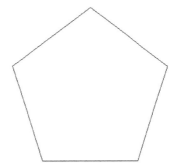

- Press Esc to deactivate the tool.

You can also enter the length, angle, and width values in the **Size A**, **Angle** and **Size B** boxes respectively to create the rotated rectangle.

Drawing Polygons

A Polygon is a single object having many sides ranging from 3 to 1024. In TurboCAD, you can create regular polygons having sides with equal length. There are three methods to create a polygon.

Vertex Mode The polygon will be created with its vertices touching an imaginary circle (Circumscribed circle). By default, vertex mode is selected to create a polygon.

Segment Mode The polygon will be created with its sides touching an imaginary circle (Inscribed circle).

Edge Mode The polygon will be created by defining its edges.

These methods are explained in the following examples.

Example 1 (Vertex Mode)
In this example, you will create a polygon by specifying the number of sides and then defining the centerpoint of the polygon, next, by defining the radius of an imaginary circle (Circumscribed circle).

- Click **Line** drop-down > **Polygon** on the **Draw** toolbar.

- Right click and select **Vertex mode** from the local menu.
- Click in the graphics window to specify the center point.
- Press the TAB key and type **5** in the **Sides** box.
- Press the TAB key and enter **18** in the **Angle** box. The orientation of the polygon is defined.
- Press the TAB key and enter **100** in the **Radius** box. The radius of the imaginary circle touching the vertices of the polygon is specified.
- Press Enter to create the polygon.

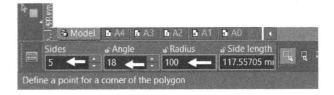

- Press Esc to deactivate the tool.

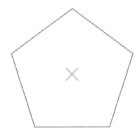

Example 2 (Segment Mode)

In this example, you will create a polygon by specifying the number of sides and then by defining the center point of the polygon, next, by defining the radius of an imaginary circle (Inscribed circle).

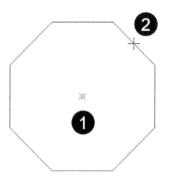

Example 3 (Edge Mode)

In this example, you will create a polygon by specifying the number of sides and then by defining the length of one side of the polygon.

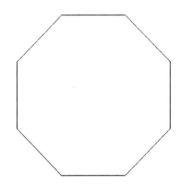

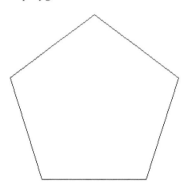

- Click **Line** drop-down > **Polygon** on the **Draw** toolbar.
- Right click and select **Segment Mode** from the local menu.

- Click **Line** drop-down > **Polygon** on the **Draw** toolbar
- Right click and select **Edge** from the local menu.

- Press the TAB key and type **8** in the **Sides** box
- Select an arbitrary point in the graphics window to define the center point of the polygon.
- Move the pointer outward to define the radius of the Inscribed circle.
- Click to create the polygon.

- Select an arbitrary point in the graphics window to define one of the corners of a polygon.
- Press the TAB key and type **5** in the **Sides** box
- Press the TAB key and enter 0 in the **Angle** box.
- Press the TAB key twice and enter 100 in the **Side Length** box.

TurboCAD 2019 For Beginners

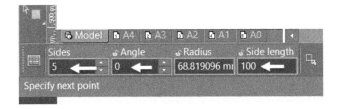

- Press ENTER to create the polygon.

Drawing Polylines

A Polyline is a single object that consists of line segments and arcs. It is more versatile than a line, as you can assign a width to it. In the following example, you will create a closed polyline.

Example 1

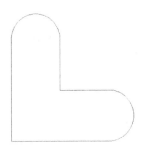

- Click the **Ortho** icon on the **Snap Modes** toolbar (or) click **Modes > Snaps > Ortho** on the Menu bar.

- Activate the **SNAP** option on the status bar, if not already active.

- Click **Line** drop-down > **Polyline** on the **Draw** toolbar.
- Click the X box next to the GEO option on the status bar.

- Type 0 and press TAB.
- Type 0 and press ENTER.

- Move the pointer horizontally toward the right.
- Press the TAB key and type 100 — next, press the ENTER key.
- Right click and select **Arc Segment** from the Local menu.
- Click the X box next to the GEO option on the status bar.

- Type 100 and press TAB.
- Type 50 and press ENTER.
- Right click and select **Line Segment** option.

- Move the pointer horizontally toward left.
- Press the TAB key and type **50** — next, press ENTER.
- Move the pointer vertically upward
- Press the TAB key and type **50** — next, press ENTER.
- Right click and select **Arc Segment** from the Local menu.
- Click the X box next to the GEO option on the status bar.

20 | Drawing Basics

TurboCAD 2019 For Beginners

- Type 0 and press TAB.
- Type 100, and press ENTER.
- Right-click and select the **Close** option from the local menu.

Now, when you select a line segment from the sketch, the whole sketch will be chosen. It is because the polyline created is a single object.

Drawing Circles

The tools in the **Circle** drop-down on the **Draw** toolbar can be used to draw circles. There are various methods to create circles. These methods are explained in the following examples.

Example 1(Center, Radius)
In this example, you will create a circle by specifying its center and radius value.

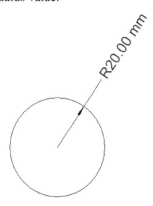

- Click **Circle/Ellipse** drop-down > **Center, Radius** on the **Draw** toolbar.

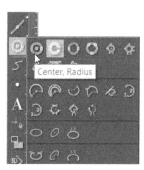

- Click in the graphics window to define the center point of the circle.
- Move the pointer outward.
- Press the TAB key.
- Type 20 in the **Radius** box and press ENTER.

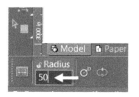

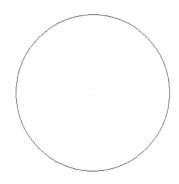

Example 2(Center, Diameter)
In this example, you will create a circle by specifying its center and diameter value.

- Click **Circle/Ellipse** drop-down > **Center Radius** on the **Draw** toolbar.
- Click in the graphics window to define the center point of the circle.
- Right click and select **Diameter** from the local menu.

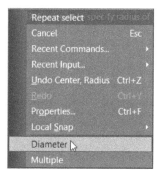

- Move the pointer outward.
- Press the TAB key.
- Type 30 in the **Diameter** box and press ENTER.

Example 3 (2 Point)

In this example, you will create a circle by specifying two points. The first point is to specify the location of the circle, and the second defines the diameter.

- Create a triangle, as shown in the figure below. Assume the dimensions.

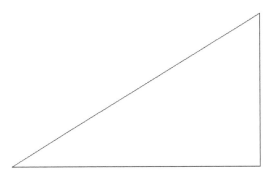

- Click **Circle/Ellipse > 2 Point** on the **Draw** toolbar.
- Select the left endpoint of the triangle.
- Move the pointer and select the top-right endpoint of the triangle; the circle will be created as shown below.

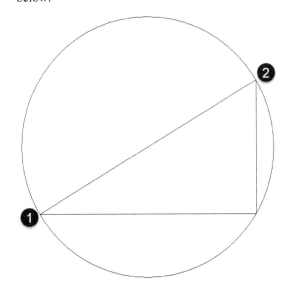

Example 4 (Circle, Tan, Tan, Tan)

In this example, you will create a circle by selecting three lines. The circle will be tangent to the three lines.

- Click **Circle/Ellipse > Circle, Tan, Tan, Tan** on the **Draw** toolbar.

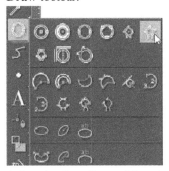

- Select the three lines of the triangle, as shown.

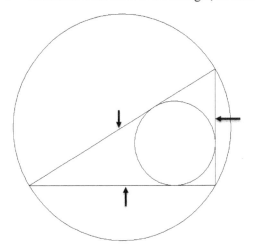

Example 5 (3 Point)

In this example, you will create a circle by specifying three points. The circle will pass through these three points.

- Open a new file.
- Click **Line** drop-down > **Polygon** on the **Draw** toolbar.
- Specify the center point of the polygon.
- Move the pointer outward.
- Press the TAB key and type 3 in the **Sides** box.
- Specify the Angle and Side Length values, as shown.
- Press ENTER to create the polygon.

- Click **Circle/Ellipse** drop-down > **3 Points** on the

ribbon.

- Select the three vertices of the triangle; a circle will be created, passing through the selected points.

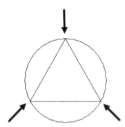

Example 6 (Concentric Circles)
In this example, you will create a series of circles concentric to each other.

- Open a new file.
- Click **Circle/Ellipse** drop-down > **Concentric Circles** on the **Draw** toolbar.
- Specify an arbitrary point in the graphics window to define the center point of the circle.
- Move the pointer and click to define the radius or diameter of the circle.

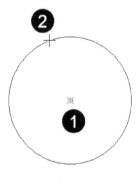

- Move the pointer outward.
- Click to create another circle concentric to the first circle.

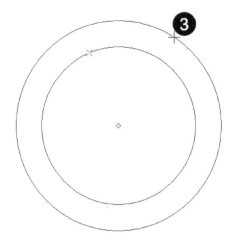

- Next, move the pointer inward and click to create another concentric circle, as shown.

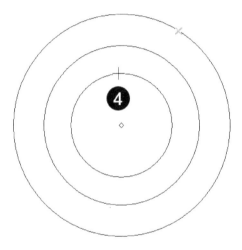

- Press Esc to deactivate the tool.

Drawing Arcs

An arc is a portion of a circle. The total angle of an arc will always be less than 360 degrees, whereas the total angle of a circle is 360 degrees. TurboCAD provides you with ten ways to draw an arc. You can draw arcs in different ways by using the tools available in the **Arc** drop-down of the **Draw 2D** panel. The usage of these tools will depend on your requirement. Some methods to create arcs are explained in the following examples.

Example 1 (Arc Start /End / Included)
In this example, you will create an arc by specifying three points. The arc will pass through these points.

TurboCAD 2019 For Beginners

- Click **Cricle** drop-down > **Arc** > **Start / End / Included** on the ribbon.

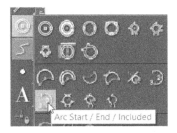

- Specify an arbitrary point in the graphics window to define the start point of an arc.
- Pick another point to define the endpoint of the arc.

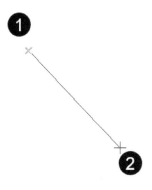

- Pick a point to define the included angle of the arc.

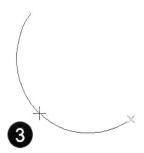

Example 2 (Arc Start / Included / End)
In this example, you will draw an arc by specifying three points. The first two points define the direction of the arc, and the third point defines its radius.

- Click **Circle drop-down > Arc Start/Included/End** on the **Draw** toolbar.

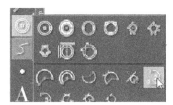

- Specify an arbitrary point in the graphics window to define the start point of an arc.

- Move the pointer and click to define the direction of the arc.

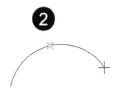

- Next, pick a point to define the radius of the arc.

Example 3 (Center and Radius)
In this example, you will draw an arc by specifying its center point and defining a point on the circumference of the circle. Then, by defining the start and end angles.

- Click **Circle** drop-down > **Arc Center and Radius** on **Draw** toolbar.

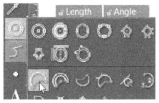

- Specify an arbitrary point in the graphics window to define the center point of an arc.
- Move the pointer outward and click to define the radius of the arc

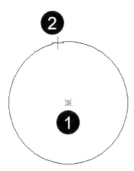

A dotted line appears from the center point of the circle.

- Move the pointer and click to define the start angle of an arc.

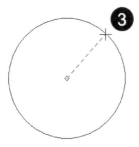

- Move the pointer in the anticlockwise direction and click to define the endpoint of an arc. (You can also move the pointer in the clockwise direction to define the end angle of an arc.)

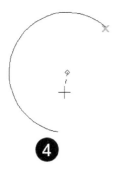

Example 4 (Concentric Arcs)

In this example, you will draw an arc by specifying its center point and defining its center, start, and endpoints. Next, create multiple arcs by using the same center point.

- Click **Circle** drop-down > **Concentric Arcs** on **Draw** toolbar.

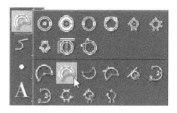

- Specify the center point of an arc.
- Move the pointer outward and click to the radius of the first arc.

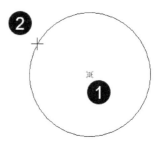

- Specify the start point of the arc.

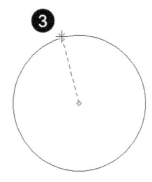

- Specify the endpoint of the arc

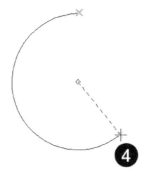

- Move the pointer outward and click to the radius of

the second arc.
- Specify the start and endpoints of the second arc.

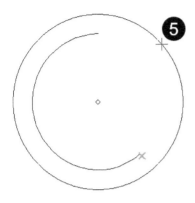

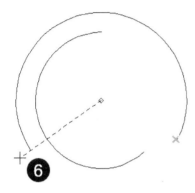

- Likewise, create as many concentric arcs as you need by sharing the same center point.

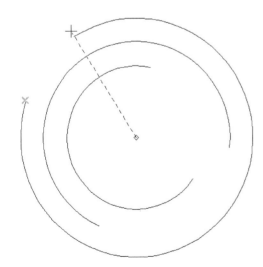

Example 5 (Double-Point)

In this example, you will draw an arc by first specifying its diameter using two points. Next, you will specify the start and endpoints of the arc

- Click **Circle** drop-down > **Double Point Arc** on **Draw** toolbar.

- Specify two points to define the diameter of the arc.

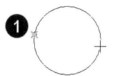

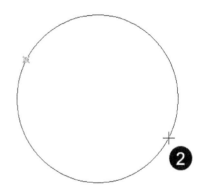

A dotted line appears from the center of the circle.
- Move the pointer and click to define the start point of the arc.

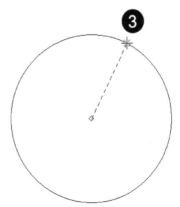

- Move the pointer in the anti-clockwise direction and click to specify the endpoint.

Example 6 (Arc Tan to Line)
In this example, you will draw an arc tangent to a line.

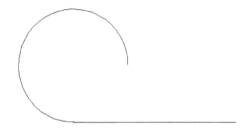

- Click **Circle** drop-down > **Arc Tan to Line** on **Draw** toolbar.

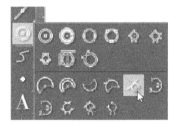

- Select an existing line to which the arc is to be tangent.

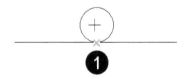

- Move the pointer and click to define the diameter of the arc.

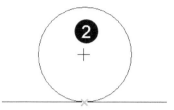

You can move the circle on either side of the line.

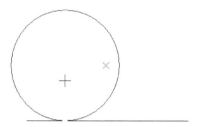

- Click to position the circle. A dotted line appears from the center of the circle.

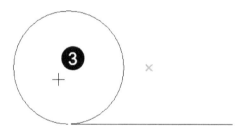

- Move the pointer and click to define the start point of the arc.

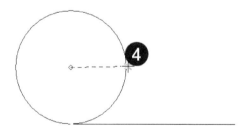

- Move the pointer in the anti-clockwise direction and click to define the endpoint of the arc; it creates an arc tangent to a line.

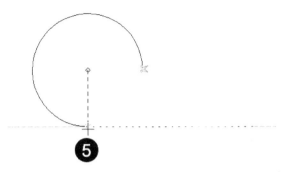

Drawing Splines

Splines are non-uniform curves, which are used to create irregular shapes. In TurboCAD, you can create splines by using three methods: **Spline By Fit Points**, **Spline By Control Points,** and **Bezier**. These methods are explained in the following examples:

Example 1: (Spline By Fit Points)

In this example, you will create a spline using the **Spline By Fit Point** method. In this method, you need to specify various points in the graphics window. The spline will be created, passing through the specified points. The Bezier method is similar to the **Spline By Fit Points** method except that the algorithm will be different.

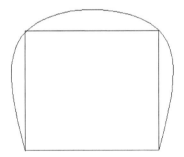

- Start a new drawing file.
- Use the **Rectangle** tool and create a sketch similar to the one shown below.

- Click **Splines > Spline By Fit Points** on the **Draw** toolbar.

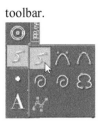

- Select the lower-left corner of the rectangle.

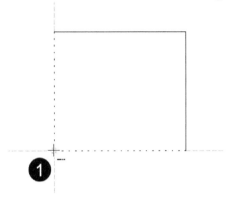

- Select the top-left corner point of the rectangle.

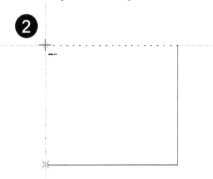

- Similarly, select the top-right and lower-right corners; a spline will be attached to the pointer.

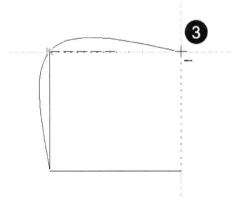

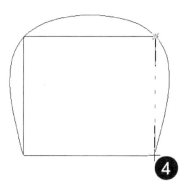

- Right-click and select **Finish** from the menu.

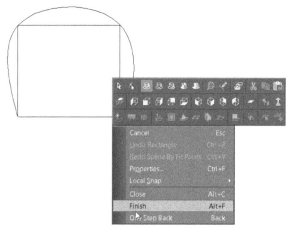

Example 2: (Spline By Control Points)

In this example, you will create a spline by using the **Spline By Control Points** method. In this method, you will specify various points called control vertices. As you specify the control vertices, imaginary lines are created connecting them. The spline will be drawn tangent to these lines.

- Click **Splines > Spline By Control Points** on the **Draw** toolbar.

- Select the four corners of the sketch in the same sequence as in the earlier example.

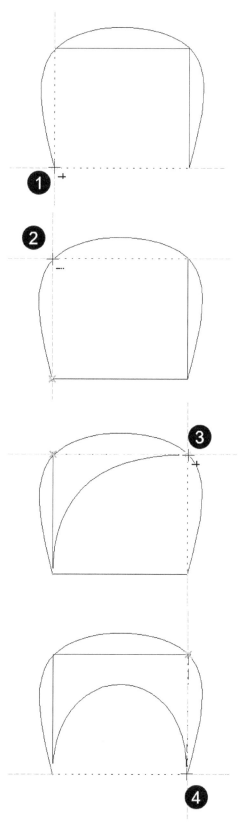

- Right-click and select **Finish** from the menu; a spline with control points will be created.

Drawing Ellipses

Ellipses are also non-uniform curves, but they have a regular shape. They are actually splines created in proper closed shape. In TurboCAD, you can draw an ellipse in three different ways by using the tools available in the **Circle/ Ellipse** drop-down of the **Draw** panel. The three different ways to draw ellipses are explained in the following examples.

Example 1 (Two corners)

In this example, you will draw an ellipse by specifying two corners.

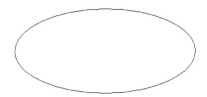

- Click **Circle / Ellipse > Ellipse** on the **Draw** toolbar.
- Select an arbitrary point in the graphics window to define the first corner of the ellipse.
- Move the pointer diagonally and click to define the second corner of the ellipse (or) enter values in the **Major** and **Minor** boxes.

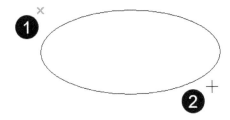

Example 2 (Rotated Ellipse)

In this example, you will draw an ellipse by specifying three points. The first two points define the angle and length of the first axis. The third point defines the second axis of the ellipse.

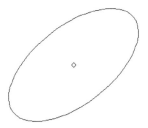

- Click **Circle / Ellipse > Rotated Ellipse** on the **Draw** toolbar.
- Select an arbitrary point to define the centerpoint of the ellipse.
- Move the mouse pointer and click to define the angle and Major axis of the ellipse (or) enter values in the **Major** and **Angle** boxes, and then click.

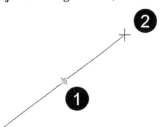

- Move the mouse pointer and click to define the second axis of the ellipse (or) enter a value in the **Minor** box; the ellipse will be created with an inclined angle.

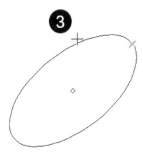

Example 3 (Ellipse Fixed Ratio)

In this example, you will draw an ellipse by specifying the ratio between the major and minor axis.

- Click **Circle / Ellipse > Ellipse Fixed Ratio** on the **Draw** toolbar.

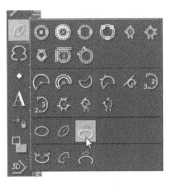

- Press the TAB key.
- Enter a value in the **a:b Ratio** box.

TurboCAD 2019 For Beginners

- Select an arbitrary point to define the centerpoint of the ellipse.
- Move the pointer outward and click to define the size of the ellipse.

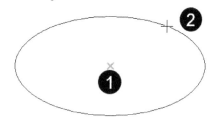

Example 4 (Elliptical Arc)

In this example, you will draw an elliptical arc. To draw an elliptical arc, first, you need to define the location and length of the first axis. Next, set the radius of the second axis; an ellipse will be displayed. Next, you need to set the start angle of the elliptical arc. The start angle can be any angle between 0 and 360. After defining the start angle, you need to specify the end angle of the elliptical arc.

- Click **Circle / Ellipse > Elliptical Arc** on the **Draw** toolbar.

- Click to specify the first corner of the ellipse.
- Move the mouse pointer diagonally and click to specify the second corner of the ellipse.

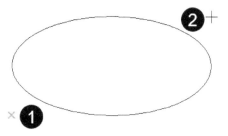

A dotted line appears from the centerpoint of the ellipse.

- Move the mouse pointer and click to define the start angle of an arc.

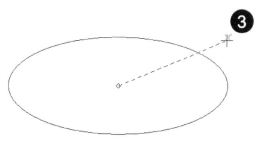

- Likewise, move the mouse pointer anti-clockwise and click to define the end angle of an elliptical arc.

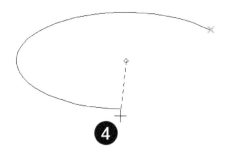

Exercises

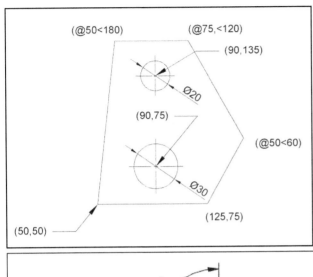

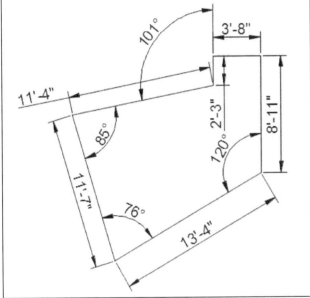

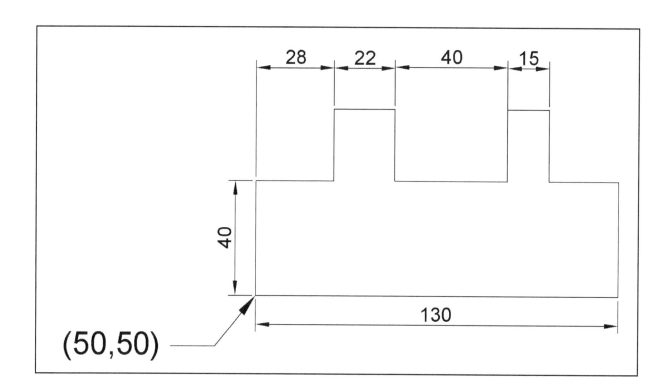

Chapter 3: Drawing Aids

In this chapter, you will learn to do the following:

- Use Grid and Snap
- Use Ortho Mode and Polar Tracking
- Use Snaps and Snap Tracking
- Create Layers and assign properties to it
- Zoom and Pan drawings

Drawing Aids

This chapter teaches you to define the drawing settings, which will assist you in creating a drawing in TurboCAD quickly. Most drawing settings can be turned on or off from the **Drawing Setup** dialog. You can open the **Drawing Setup** dialog by clicking **Options > Drawing Setup** on the Menu bar.

Setting Grid and Snap

Grid is the primary drawing setting. It makes the graphics window appear like a graph paper. You can turn ON the grid display by clicking the **Grid** icon on the Snap **Modes** toolbar or just pressing **Shift +Alt+ G** on the keyboard.

SNAP is used for drawing objects by using the intersection points of the grid lines. When you turn the SNAP Mode ON, you will be able to select only grid points. In the following example, you will learn to set the grid and snap settings.

Example:

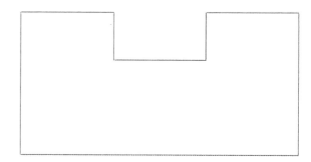

- Click **File > New** on the ribbon; the **New TurboCAD 2019 Drawing** dialog appears.
- Click the **New from Template** on the dialog; the **New from Template** dialog appears.
- Select the **ISO** template and click **OK**.
- On the Menu bar, click **Options > Drawing Setup**; the **Drawing Setup** dialog appears.
- Click the **Grid** option on the left side of the dialog.
- Enter **10** in the **X** and **Y** boxes under the **Spacing** section.

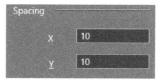

- Select the **Ortho** option from the **Type** section.
- Select the **Lines** option from the **Styles** section.

- Click **OK** on the dialog.
- Activate the **Grid** icon on the **Snap Modes** toolbar.
- Click the **Show/Hide** icon on the **Main** toolbar.

- Click the **SNAP** icon on the status bar.

35 | Drawing Aids

Next, you need to change the line width. Line width is the thickness of the objects that you draw. In TurboCAD, there is a default lineweight assigned to objects. However, you can change the line width. The method to set the lineweight is explained below.

- Click **Line** drop-down > **Rectangle** on the **Draw** toolbar.
- Select an arbitrary point from the graphics window.
- Press the TAB key.
- Type **80** in the **Size A** box and press the TAB key.
- Type **50** in the **Size B** box and press ENTER.

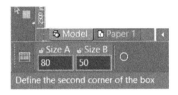

- Click the **Extents** icon on the **Zoom** toolbar. The area inside the rectangle is zoomed in.
- Click the **Line** icon on the **Draw** toolbar.
- On the **Property** toolbar, click in the **Line Width** box and enter **0.5**.

- Type **L** in the command line and press ENTER.
- Select the lower left grid point inside the rectangle, as shown.

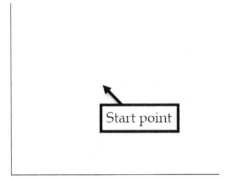

- Move the pointer horizontally toward the right and click on the sixth grid point from the first point.
- Move the pointer vertically upwards and select the third grid point from the second point.
- Move the pointer horizontally toward the left and select the second grid point from the previous point.
- Move the pointer vertically downwards and select the grid point next to the previous point.
- Move the pointer horizontally toward the left and select the second grid point from the previous point.
- Move the pointer vertically upwards and select the grid point next to the previous point.
- Move the pointer horizontally toward the left and select the second grid point from the previous point.
- Right-click and select **Close**.

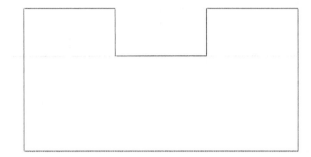

- Save and close the file.

Specifying the Angle for Ortho mode

Ortho mode is used to draw orthogonal (horizontal or vertical) lines. However, you can change the ortho angle to constrain the lines to angular increments. In the following example, you will create a drawing with the help of Ortho Mode.

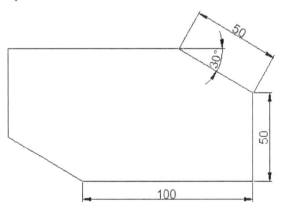

- Open a new TurboCAD file.
- Deactivate the **Grid** and **Show/Hide Grid** icons.
- Click the **Ortho** icon on the Snape Modes toolbar.
- Click the **Extents** icon on the **Zoom** toolbar.

TurboCAD 2019 For Beginners

- On the menu bar, click **Options > Drawing Sub-Options > Angle**.

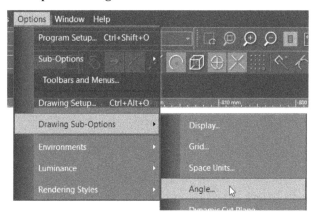

- Enter **30** in the **Step angle** box.
- Click **OK** on the **Drawing Setup** dialog.
- Click the **Line** button on the **Draw** toolbar.
- Select an arbitrary point to define the starting point.
- Move the pointer toward the right.
- Press the TAB key, type 100, and press ENTER; you will notice that a horizontal line is created.
- Move the pointer upwards.
- Press the TAB key, type 50, and press ENTER; you will notice that a vertical line is created.
- Move the pointer and stop when the **Angle** box displays 150.

- Press the TAB key, type 50, and press ENTER.

- Move the pointer toward left.
- Press the TAB key.
- Type 100 and press ENTER when the **Angle** box displays 180.

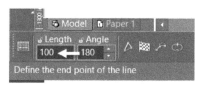

- Move the pointer vertically downward.
- Press the TAB key.
- Type 50 and press ENTER when the **Angle** box displays 270.
- Right-click and select **Close**.

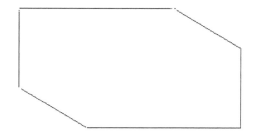

Using Layers

Layers are like a group of transparent sheets that are combined into a complete drawing. The figure below displays a drawing consisting of object lines and dimension lines. In this example, the object lines are created on the 'Object' layer, and dimensions are created on the layer called 'Dimension.' You can easily turn-off the 'Dimension' layer for a more unobstructed view of the object lines.

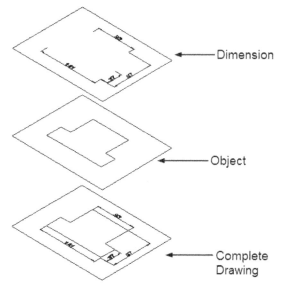

Layer Manager

37 | **Drawing Aids**

The **Layer Manager** is used to create and manage layers. To open **Layer Manager**, click **Format > Layers** on the menu bar or click **Layers** icon on the **Layers** toolbar.

The components of the **Layer Manager** are shown below. The **Tree View** section is used for displaying layer filters, group, or state information. The **List View** section is the main body of the **Layer Manager**. It lists the individual layers that currently exist in the drawing.

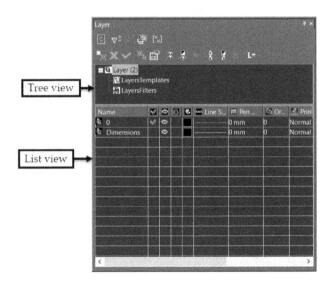

The **List View** section contains various properties. You can set layer properties and perform multiple operations in the **List View** section. A brief explanation of each layer property is given below.

Active – Shows a green check when a layer is set to current.
Name - Shows the name of the layer.
Visible – It is used to turn on/off the visibility of a layer. When a layer is turned on, it shows an eye symbol. When you turn off a layer, the eye symbol disappears.
Lock - It is used to lock the layer so that the objects on it cannot be modified.
Color – It is used to assign a color to the layer.
Line Style – It is used to assign a line style to the layer.
Pen Width – It is used to define the line width (thickness) of objects on the layer.
Order – It is used to move the layer to the front or back.

Print Style – It is used to override the settings such as color, linetype, and lineweight while plotting a drawing.
Plot – It is used to control which layer will be plotted.
VP Visible – It is used to hide or display a layer in any viewport.

Creating a New Layer
You can create a new layer by using any one of the following methods:

1. Click the **New Layer** button on the **Layer Manager**; the **New Layer** dialog appears. Next, enter the name of the layer in the **New Layer** dialog, and then click **OK**.

2. Right-click in the **Name** field and select **New Layer** from the local menu.

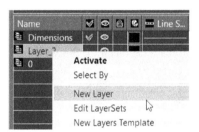

Making a layer current
If you want to draw objects on a particular layer, then you have to make it current. You can make a layer current using the methods listed below.

1. Select the layer from the List view and click the **Activate** button on the **Layer Manager**.

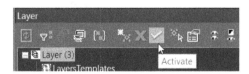

2. Double-click on the **Name** field of the layer.
3. Right-click on the layer and select **Activate**.
4. Select the layer from the **Layer** drop-down of the **Layer** toolbar.

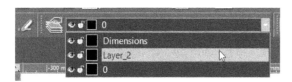

Deleting a Layer

You can delete a layer by using any one of the following methods:

1. Click the **Delete Layer** button.

2. Right-click in the **Name** field and select **Delete Layer** from the local menu.

Using Local Snaps

Local Snaps are essential settings that improve your performance and accuracy while creating a drawing. They allow you to select key points of objects while creating a drawing. You can activate the required Local Snaps by using the **Local Snaps** local menu. Right-click and place the pointer on the **Local Snap** option to display this local menu. Note that the local snaps can be used only when a drawing command is active.

The functions of many Local Snaps are explained next.

Vertex: Snaps to points of dimension lines, text objects, dimension text, and an endpoint of the line.

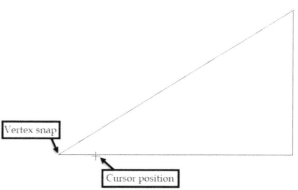

Middle point: Activate this option and click on a line; the midpoint of the line is selected.

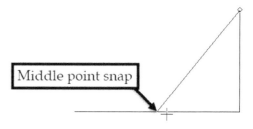

Intersection: Select this option and click near the intersection of two lines; the intersection point is selected.

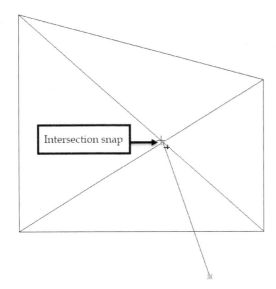

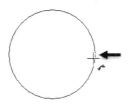

Tangent: Snaps to the tangent points of arcs and circles.

Apparent Intersection: Snaps to the projected intersection of two objects in 3D space.

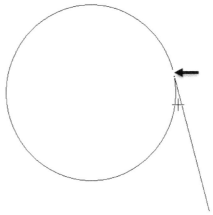

Projection: Snaps to a perpendicular location on an object.

Center: Snaps to the centers of circles and arcs.

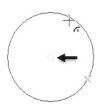

Nearest On Graphic: Snaps to the nearest point found along with any object.

Quadrant Point: Snaps to four key points located on a circle.

No Snap: Deactivates the Object Snap.

Perform 2 Mid points Snap: Snaps to the middle point of two selected points.

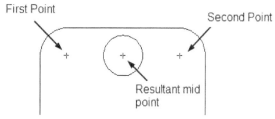

Running Snap Modes

Previously, you have learned to select snaps from the **Local Snap** local menu. However, you can make Snap modes available continuously instead of picking them every time. You can do this by using the Running **Snap Modes**. To use the Running Snap Modes, click **Modes > Snaps** on the menu bar and select the required object snap from the menu.

You can also select the **Snap Options** option from the menu to open the **Drawing Aids** dialog. In this dialog, you can choose the required **Running Snaps** by selecting checkboxes.

TurboCAD 2019 For Beginners

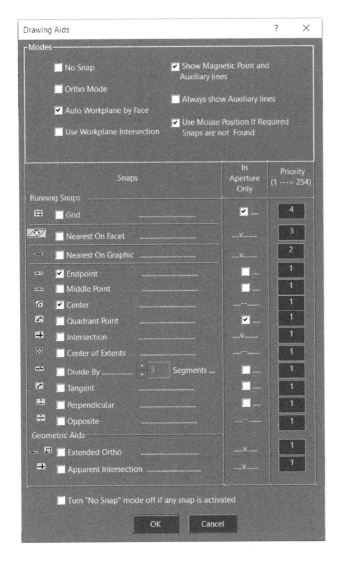

- Click **OK**.

Using Zoom tools

Using the zoom tools, you can magnify or reduce a drawing. You can use these tools to view the minute details of a very complicated drawing. The Zoom tools can be accessed from the **Zoom** toolbar and Menu Bar.

Using Auxiliary Lines

The Auxiliary Lines originate from the magnetic points of the objects. The magnetic points and the auxiliary lines are displayed only when the **Show Magnetic Point** icon is activated on the **Snap Modes** toolbar (or) check the **Show Magnetic Points and Auxiliary Lines** option on the **Drawing Aids** dialog.

- Select the **Show Magnetic Point** icon from the **Snap Modes** toolbar.

(OR)

- Click **Modes > Snaps > Snap Options** on the menu bar.
- Check the **Show Magnetic Points and Auxiliary Lines** option.

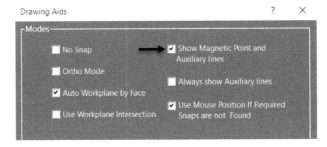

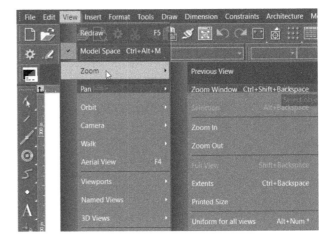

Zooming with the Mouse Wheel

Zooming using the mouse wheel is one of the easiest methods.

Drawing Aids

- Roll the mouse wheel forward to zoom into a drawing.
- Roll the mouse wheel backward to zoom out of the drawing.
- Press the mouse wheel and drag the mouse to pan the drawing.

Using Zoom Extents

Using the **Zoom Extents** tool, you can zoom to the extents of the largest object in a drawing.

- Click **Zoom Extents** on the **Zoom** toolbar.
- You can also click on the mouse wheel to zoom to extents.

Using Zoom-Window

Using the **Zoom-Window** tool, you can define the area to be zoomed by selecting two points representing a rectangle.

- Click **Zoom Window** on the **Zoom** toolbar.
- Specify the first point of the zoom window, as shown.
- Move the pointer diagonally toward the right, and then specify the second point, as shown.
  The area inside the window will be zoomed.

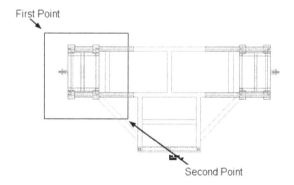

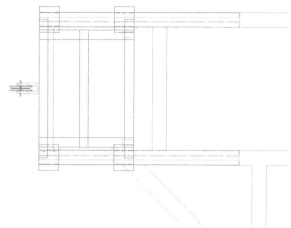

Using the Previous View

After magnifying a small area of the drawing, you can use the **Previous View** tool to return to the previous display.

- Click **View > Zoom > Previous View** on the menu bar.

Using Zoom Selection

Using the **Zoom Selection** tool, you can magnify a portion of the drawing by selecting one or more objects.

- Select one or more objects from the drawing.
- Click **Zoom Selection** on the **Zoom** toolbar; the objects will be magnified.

Using Zoom In

Using the **Zoom In** tool, you can magnify the drawing by a scale factor of 2.

- Click **Zoom In** on the **Zoom** toolbar; the drawing is magnified to double.

Using Zoom Out

The **Zoom out** tool is used to de-magnify the display screen by a scale factor of 0.5.

Panning Drawings

After zooming into a drawing, you may want to view an area that is outside the current display. You can do this by using the **Pan** tool.

- Click **View > Pan > Point to Point**.
- Click in the graphics area to pan to that area.
- Click **View > Pan > Vector Pan**.
- Pick a point in the graphics window.
- Move the pointer and click again; the drawing is panned to the area in which the point is specified.

Exercises

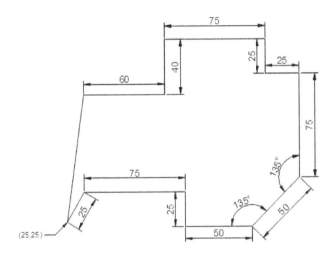

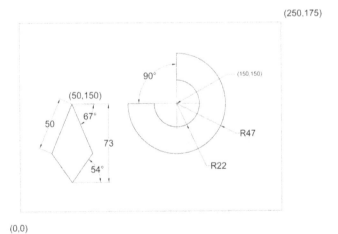

Chapter 4: Editing Tools

In this chapter, you will learn the following tools:

- The **Move** tool
- The **Vector Copy** tools
- The **Rotate** tool
- The **Scale** tool
- The **Trim** tool
- The **Shrink/Extend** tool
- The **Fillet 2D** tool
- The **Chamfer** tool
- The **Mirror Copy** tool
- The **Explode** tool
- The **Stretch** tool
- The **Radial Copy** tool
- The **Offset** tool
- The **Array** tool
- Grips

Editing Tools

In previous chapters, you have learned to create some simple drawings using the basic drawing tools. However, to create complex drawings, you may need to perform various editing operations. The tools to perform the editing operations are available on the **Modify** toolbar and the **Modify** menu. You can click the down arrow on this panel to find more editing tools. Using these editing tools, you can modify existing objects or use existing objects to create new or similar objects.

The Move tool

The **Move** tool moves the selected object(s) from one location to a new location without changing its orientation. To move objects, you must select the objects from the graphics window and activate this tool. After selecting the objects, you must define the 'base point' and the 'destination point.'

Example:

- Create the drawing, as shown below.

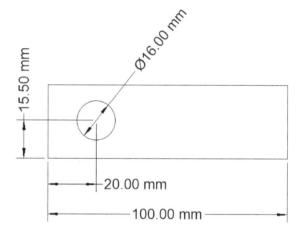

- Click the **Edit** tool on the **Draw** toolbar.

- Select the circle.
- Right click and select **Select Edit**.
- Click **Modfiy > Transform > Move** on the Menu bar (or) click **Transform** drop-down > **Move** on the **Modify** toolbar.

- Click **OK** on the **TurboCAD 2019 Warning** dialog.

TurboCAD 2019 For Beginners

- Make sure that the **Ortho** icon is activated on the **Snap Modes** toolbar.
- Activate the **Keep Original Object** icon on the Inspector Bar, if not already active.

- Select the center point of the circle.
- Move the pointer toward the right.
- Press the TAB key
- Type 30 and press ENTER.
- Click on the circle located on the right side.

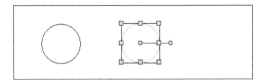

- Deactivate the **Keep Original Object** icon on the Inspector Bar.
- Select the center point of the circle to define the base point.

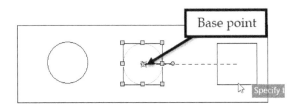

- Move the pointer toward the right.
- Press the TAB key.
- Type 30 and press ENTER; the circle moves the new location.

The Vector Copy tool

The **Vector Copy** tool is used to copy objects and place them at a required location. This tool is similar to the **Move** tool, except that object will remain at its original position, and a copy of it will be placed at the new location.

Example:

- Draw two circles of 80 mm and 140 mm diameter, respectively.

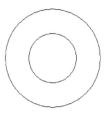

- Click the **Edit** tool on the **Draw** toolbar.
- Press the Space key.
- Create a selection box across the two circles.

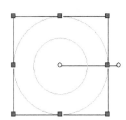

- Click **Modify > Copy > Vector** (or) click **Copy** drop-down > **Vector Copy** on the **Modify** toolbar.

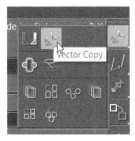

- Make sure that the **Ortho** mode is active.
- Select the center point of the circles.
- Move the pointer toward the right.
- Press the TAB key.

- Type **200** in the **Length** box; This action creates a copy of the circles at the new location.

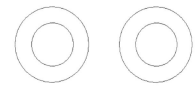

The Rotate tool

The **Rotate** tool rotates an object or a group of objects about a base point. Activate this tool and select the objects from the graphics window. After selecting objects, you must define the 'base point' and the angle of rotation. This rotates the object(s) about the base point.

- On the Menu bar, click **Tools > Select**.
- Create a selection window across the concentric circles on the left side, as shown.

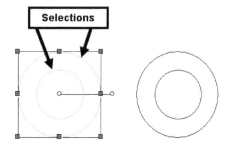

- Click **Modfiy > Transform > Rotate** on the Menu bar (or) click **Transform** drop-down > **Rotate** on the **Modify** toolbar.

- Click **OK** on the **TurboCAD 2019 Warning** dialog.
- Place the mouse pointer on the circle on the right side.
- Click on the center point of the circle as the base point, as shown.

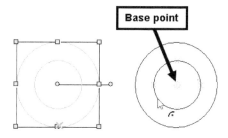

- Select the center pointer of the selected circles.
- Press the TAB key.
- Type-in **270** in the **Angle** box.

- Press ENTER; the selected objects are rotated by 270 degrees, as shown.

TurboCAD 2019 For Beginners

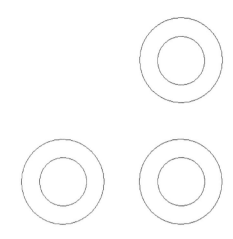

The Scale tool

The **Scale** tool changes the size of objects. It reduces or enlarges the size without changing the shape of an object.

- Create a selection window across the concentric circles on the right side, as shown.

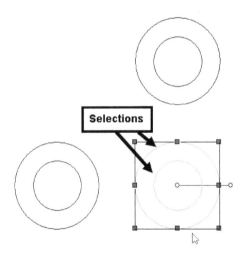

- Click **Modfiy > Transform > Scale** on the Menu bar (or) click **Transform** drop-down > **Scale** on the **Modify** toolbar.

- Right-click and deselect the **Keep Original Object** option.
- Select the center point of the selected circles as the base point.
- Select any one of the quadrant points.
- Move the mouse pointer inwards.
- Type **0.8** in the **Scale** box and press ENTER.

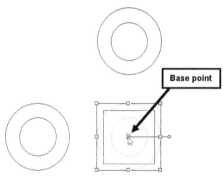

- Likewise, scale the circles located at the top to 0.7.

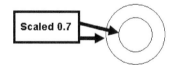

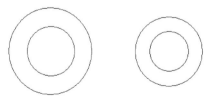

48 | Editing Tools

- Click **Draw > Circle > Tan Tan Radius** (or) click **Circle/Ellipse** drop-down > **Circle Tan Tan Rad** on the **Draw** toolbar.

- Select the two circles shown below to define the tangent points.

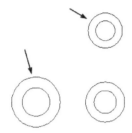

- Press the TAB key.
- Type 150 in the **Radius** box and press ENTER.

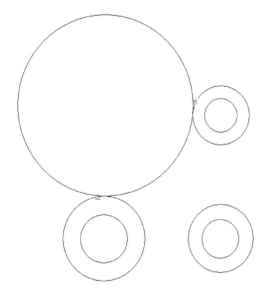

- Likewise, create other circles of radius 100 and 120.

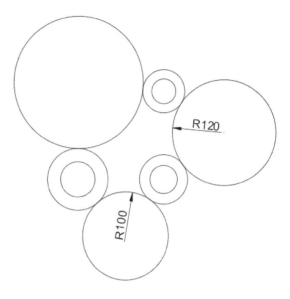

The Trim tool

When an object intersects with another object, you can remove its unwanted portion by using the **Trim** tool. To trim an object, you must first activate the **Trim** tool, and then select the cutting edge (intersecting object) and the portion to be removed. If there are multiple intersection points in a drawing, you can simply select the **Select all** option from the local menu; all the objects in the drawing objects will act as 'cutting edges.'

- Click **Modify > Modify 2D Objects > Trim** on the menu bar (or) click **Modify 2D Objects** drop-down > **Trim** on the **Modify** toolbar.

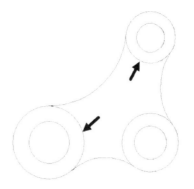

Now, you must select the cutting edges.

- Right-click and select the **Select All** option from the local menu; it selects all the objects as the cutting edges.
- Right click and select the **Finish Selection** option from the local menu.

Now, you must select the objects to be trimmed.

- Click on the portions of the large circles one by one; the circles will be trimmed.

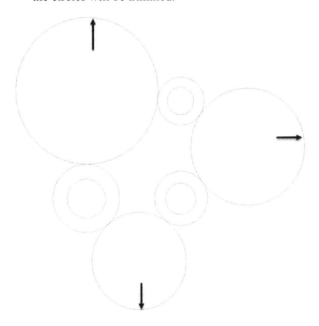

- Save and close the drawing.

The Shrink/Extend tool

The **Shrink/Extend** tool is similar to the **Trim** tool, but its use is the opposite of it. This tool is used to extend lines, arcs, and other open entities to connect to other objects. To do so, you must select the boundary up to which you want to extend the objects, and then select the objects to be extended.

- Start a new drawing.
- Create a sketch, as shown below, using the **Line** tool.

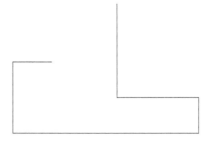

- Press Esc.
- Click **Modify 2D Objects** drop-down > **Trim** on the **Modify** toolbar.
- Right click and select the **Select All** option.
- Right click and select the **Finish Selection** option from the local menu.
- Select the inner portions of the circles, as shown.

- Click **Modify > Shrink/Extend** on the menu bar (or)

click **Modify 2D Objects** drop-down > **Shrink/Extend** on the **Modify** toolbar.

- Select the horizontal open line to change its length.
- Select the vertical line to define the new length of the line segment; this will extend the line up to the boundary line.

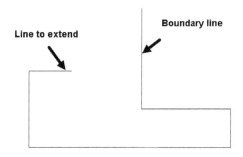

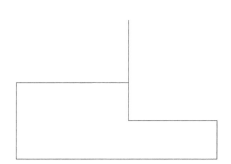

The Fillet 2D tool

The **Fillet 2D** tool converts the sharp corners into round corners. You must define the radius and select the objects forming a corner. The following figure shows some examples of rounding the corners.

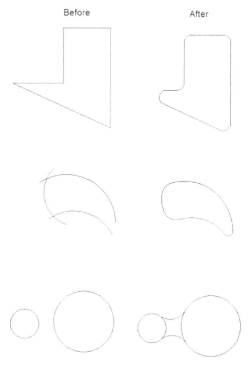

- Start a new drawing.
- Click **Zoom In** on the **Zoom** toolbar.
- Click **Line** drop-down > **Polyline** on the **Draw** toolbar.
- Draw the lines, as shown below.

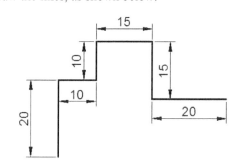

- Right-click and select **Finish**.
- Click **Modify** > **Modify 2D Objects** > **Fillet2D** on the menu bar (or) click **Modify 2D Objects** drop-down > **Fillet2D** on the **Modify** toolbar.

TurboCAD 2019 For Beginners

- Press the TAB key.
- Type **5** in the **Radius** box.

- Select the vertical and horizontal lines, as shown below.

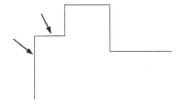

- Notice that a fillet is created.

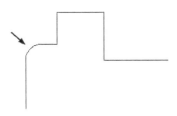

The Chamfer tool

The **Chamfer** tool replaces the sharp corners with an angled line. This tool is similar to the **Fillet 2D** tool, except that an angled line is placed at the corners instead of rounds.

- Click **Modify > Modify 2D Objects > Chamfer** drop-down >**Distance/Distance** on the menu bar (or) click **Modify 2D Objects** drop-down > **Chamfer (Distance and Distance)**.

- Press the TAB key and type 8 in the **Dist A** box.
- Press the TAB key and type 8 in the **Dist B** box.
- Press ENTER.

- Select the vertical line on the right-side.
- Select the horizontal line connected to the vertical line.

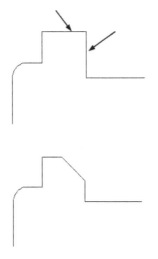

The Mirror Copy tool

The **Mirror Copy** tool creates a mirror image of objects. You can create symmetrical drawings using this tool. Select the objects to mirror and activate this tool. Next, define the 'mirror line' about which the objects will be mirrored. You can define the mirror line by either creating a line or selecting an existing line.

- Click Tools > Select on the menu bar.
- Create a selection window across the drawing on the left side.

52 | Editing Tools

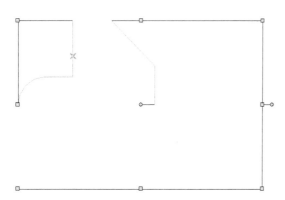

- Click **Modify > Copy > Mirror** on the menu bar (or) click **Copy** drop-down > **Mirror Copy** on the **Modify** toolbar.
- Make sure that the **Ortho** mode is activated on the **Snap Modes** toolbar.
- Select the lower-left endpoint of the polyline, as shown.
- Move the pointer toward the right and click to create the mirror line, as shown below.

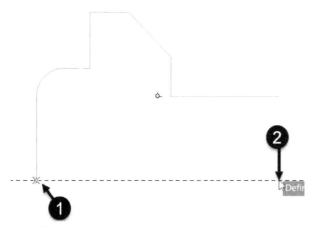

- Click anywhere in the graphics window to deselect the drawing.

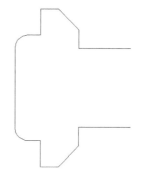

- Click the **Circle** drop-down > **Arc Tan to 2 entities** on the **Draw** toolbar.

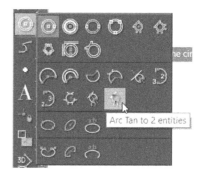

- Select the two horizontal lines, as shown.

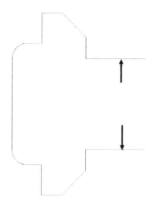

- Select the endpoint of the lower horizontal line.

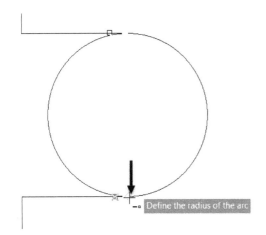

- Move the pointer toward the right and click to define the side of the arc.

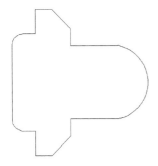

The Explode tool

The **Explode** tool explodes a group of objects into individual objects. For example, when you create a drawing using the **Polyline** tool, it acts as a single object. You can explode a polyline or rectangle or any group of objects using the **Explode** tool.

- Click the **Edit** tool on the **Draw** toolbar.
- Press and hold the CTRL key and select the two polylines.

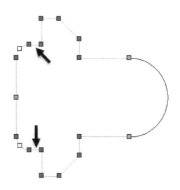

- Press the Space bar on your keyboard.
- Click **Modify > Explode** on the menu bar (or) click the **Explode** icon on the **Modify** toolbar.

The Stretch tool

The **Stretch** tool lengthens or shortens drawings or parts of drawings. Note that you cannot stretch circles using this tool. Also, you must select the portion of the drawing to be stretched by dragging a window.

- Click **Modify > Stretch** on the menu bar (or) click **Modify 2D Objects** drop-down > **Stretch** on the **Modify** toolbar.
- Create a crossing window to select the objects of the drawing.

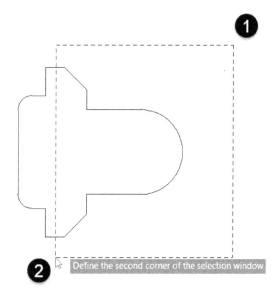

- Select the base point, as shown below.

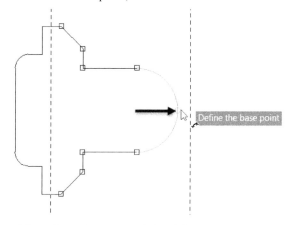

- Move the pointer toward the right.
- Press the TAB key.
- Type 10 and press ENTER to define the destination point.

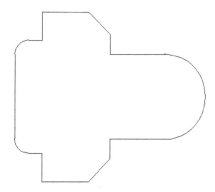

- Save and close the file.

The Radial tool

The **Radial** tool creates an array of objects around a point in a circular form. The following example shows you to create a polar array.

- Create two concentric circles of 140 and 50 diameters.

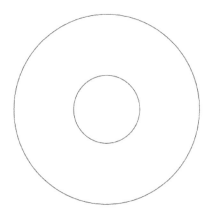

- Click **Circle** drop-down > **Center, Radius** on the **Draw** toolbar.
- Right-click and **Local Snap** > **Quadrant Point** from the local menu.
- Select the quadrant point of the circle, as shown below.

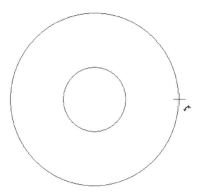

- Press the TAB key.
- Type 30 as radius and press ENTER.

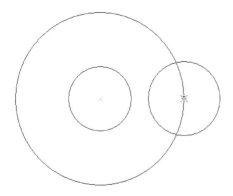
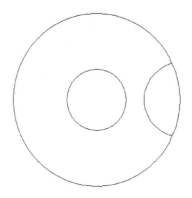

- Click **Modify > Modify 2D Objects > Object Trim** on the menu bar (or) click **Modify 2D Objects** drop-down > **Object Trim** on the **Modify** toolbar.

- Right click and select **Cancel**.
- Click the Edit tool icon on the Draw toolbar.
- Select the arc created after trimming the circle.
- Press the Space bar.

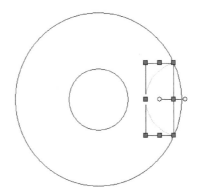

- Select the large circle to define the cutting edge.
- Select the outer portion of the small circle.

- Click **Modify > Array > Radial** on the menu bar (or) click **Copy** drop-down > **Radial Copy** on the **Modify** toolbar.

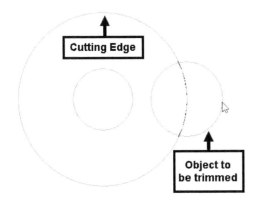

- Move the pointer and click on the inner circle; the center point of the inner circle is selected.
- Press the TAB key.
- Type-in **4** in the **Sets** box.

- Press ENTER to create an array.

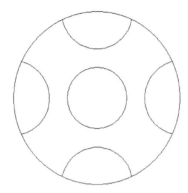

- Click **Edit > Modify 2D Objects > Trim** on the menu bar.
- Right click and select **Select All** from the local menu.
- Right click and select **Finish Selection**.
- Select the portion of the outer circle as the object to be trimmed.

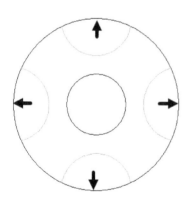

- Press Esc to deactivate the tool.

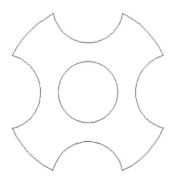

The Offset tool

The **Offset** tool creates parallel copies of lines, polylines, circles, and arcs. To create a parallel copy of an object, first, you must define the offset distance and then select the object. Next, you must define the side in which the parallel copy will be placed.

- Click the **Line** tool on the **Draw** toolbar and create a drawing, as shown.

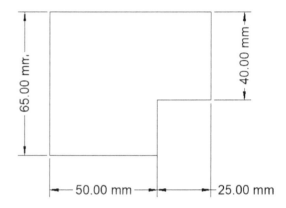

- Click **Modify > Offset** on the menu bar (or) click **Copy** drop-down > **Offset** on the **Modify** toolbar.

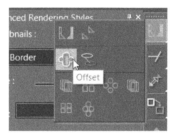

- Press the TAB key.
- Type **5** in the **Offset Distance** box and press ENTER.
- Select the top horizontal line and click inside the loop. The parallel copy of the selected is created.

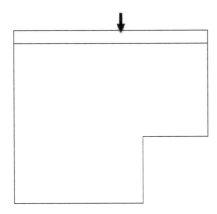

- Select the top horizontal line and click inside the loop. The parallel copy of the entire loop is created.

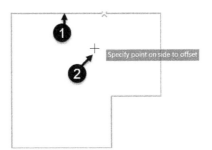

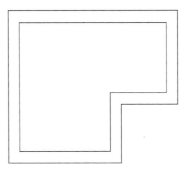

- Likewise, offset the other lines of the drawing.

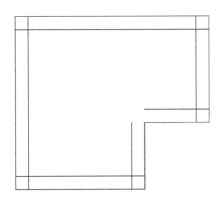

The Array tool

The **Array** tool creates an array of objects along with the X and Y directions.

- Click the **Line** drop-down > **Polyline** on the **Draw** toolbar and create a drawing, as shown.

- Open a new TurboCAD file and draw the sketch shown below. Do not add dimensions. (refer to the **Drawing Rectangles** and **Drawing Circles** section in Chapter 2 to know the procedure to draw the rectangle and circle)

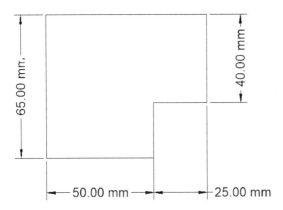

- Click **Modify** > **Offset** on the menu bar (or) click **Copy** drop-down > **Offset** on the **Modify** toolbar.
- Press the TAB key.
- Type **5** in the **Offset Distance** box and press ENTER.

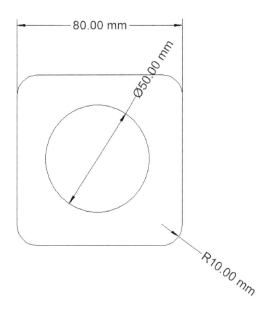

- Click **Circle** drop-down > **Center, Radius** on the **Draw** toolbar (or) click **Draw** > **Circle** > **Center, Radius** on the menu bar.
- Right click and select **Local Snap** > **Center**.
- Select the lower-left fillet.

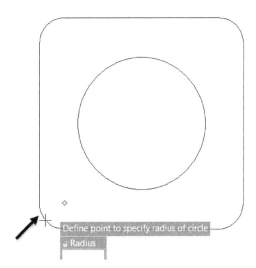

- Press the TAB key.
- Type-in **5** in the **Radius** box and press ENTER.

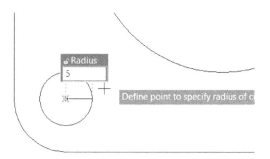

- Click the **Edit tool** icon on the Draw toolbar.
- Select the small circle.
- Press the Space bar.

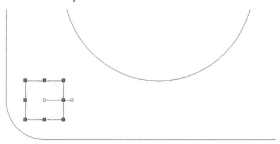

- Click **Modify** > **Array** > **Array** on the menu bar (or) click **Copy** drop-down > **Array Copy** on the **Modify** toolbar.

- Right click and select **Local Snap** > **Center**.
- Move the pointer and select the top-right fillet; the center point of the fillet is selected.

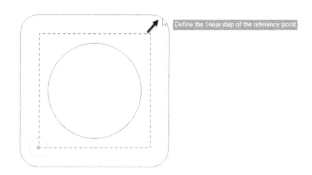

- Press Esc.

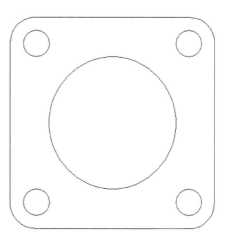

Editing Using Grips

The **Edit Tool** allows you to edit the objects using the grips. Click the **Edit Tool** icon on the **Draw** toolbar and select objects from the graphics window; small squares appear on them. These squares are called grips. You can use these grips to stretch, move, rotate, scale and mirror objects change properties, and perform other editing operations. Grips displayed on selecting different objects are shown below.

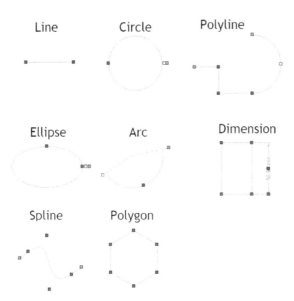

The following table gives you the details of the editing operations that can be performed when you select and drag grips.

Object	Grip	Editing Operation
Circle	Grip on circumference	**Stretch:** Select any one of the grips on the circumference and move the pointer to stretch a circle.

	Angle grip	Select this grip and move the pointer; the included angle of the circle changes.
Arc	Grip on circumference	**Resize**: Select the grip on the circumference and move the pointer.
	Endpoint Grip	**Stretch/Lengthen**: Select an endpoint grip and move the pointer.

Polylines, Rectangles, Polygons	Corner Grips	**Stretch**: Select the corner grips and move the pointer. 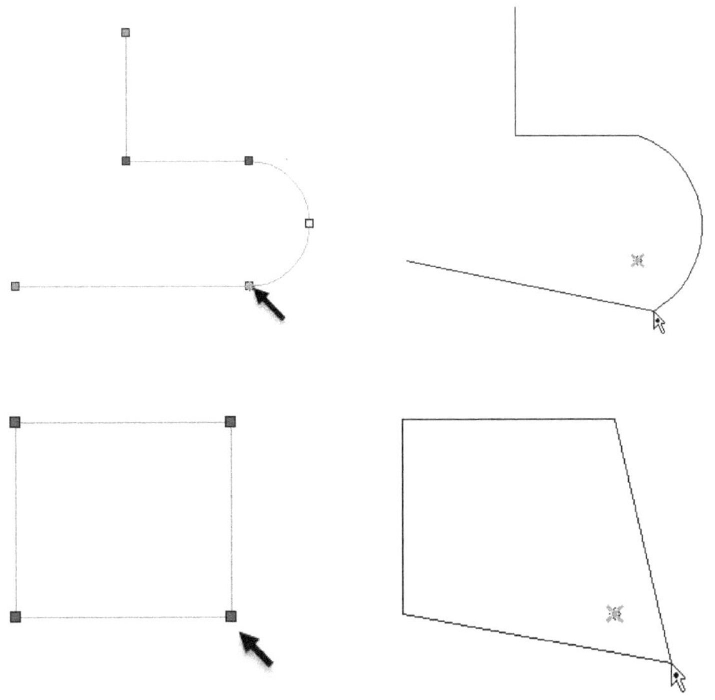
Ellipse	Grips on circumference	**Stretch**: Select a grip on the circumference and move the pointer. 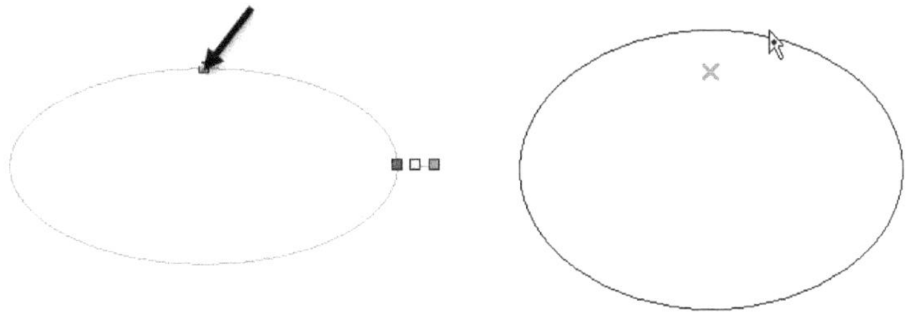
Spline	Edit Fit Points	**Stretch**: Select the spline and click the **Edit Fit Point** icon on the Inspector Bar. Next, select a grip on the spline and move the pointer.  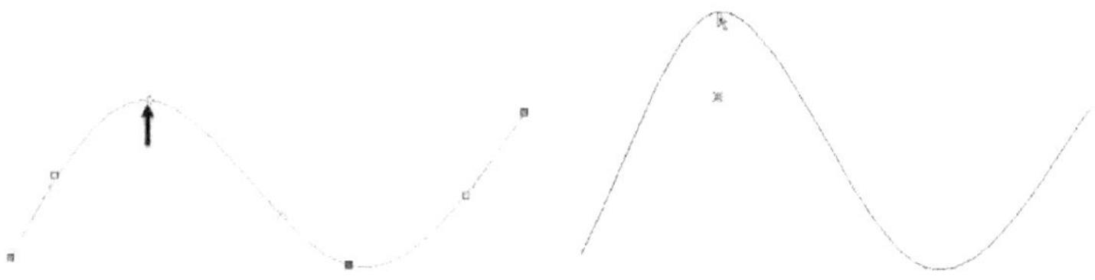

Edit Control Points	**Stretch Vertices**: Select the spline and click the **Edit Control Points** icon on Inspector Bar. Select the control vertices of the spline and move the pointer. 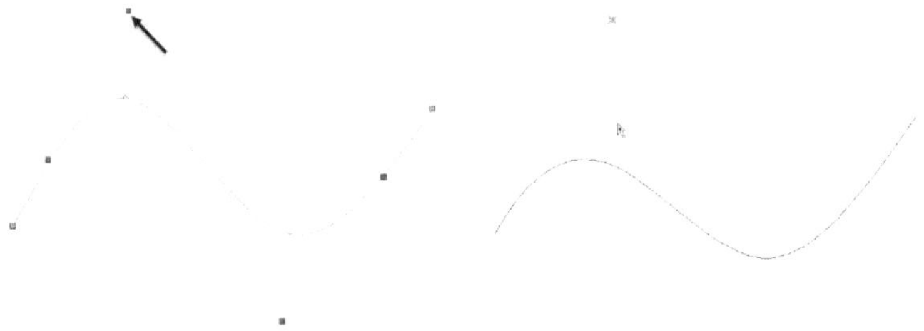

Exercises

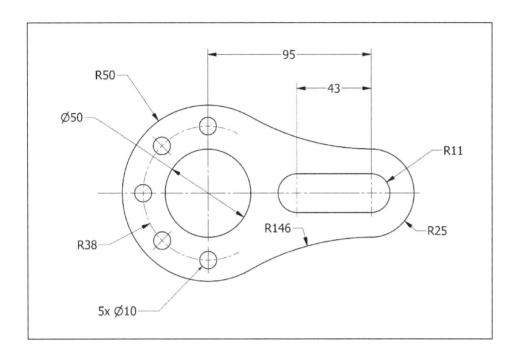

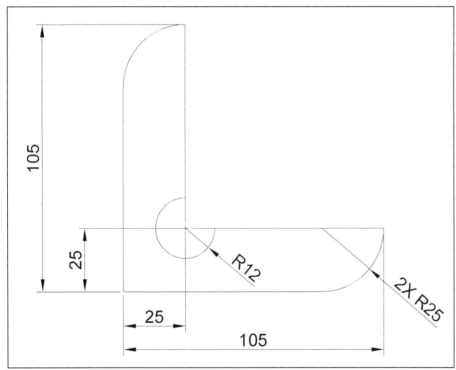

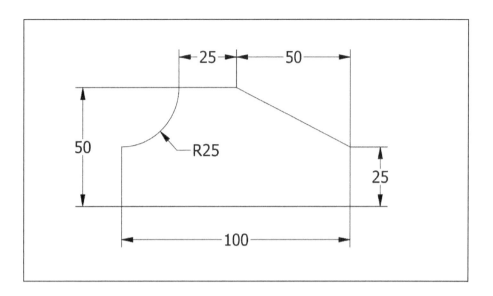

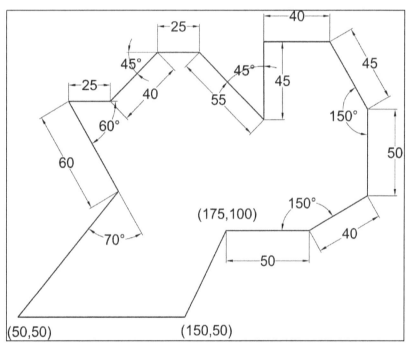

TurboCAD 2019 For Beginners

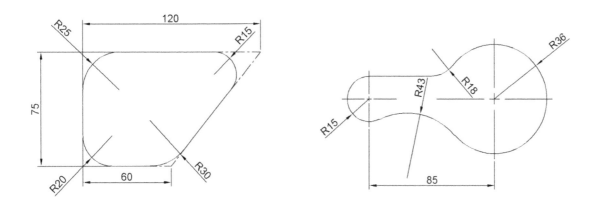

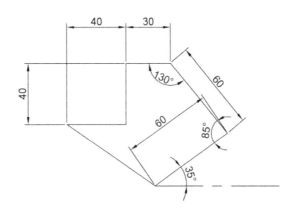

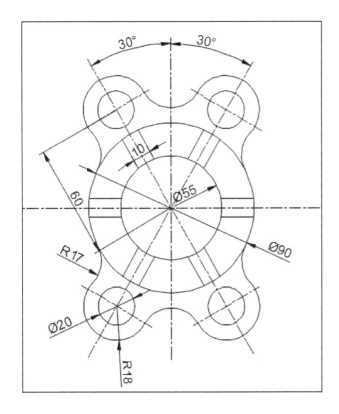

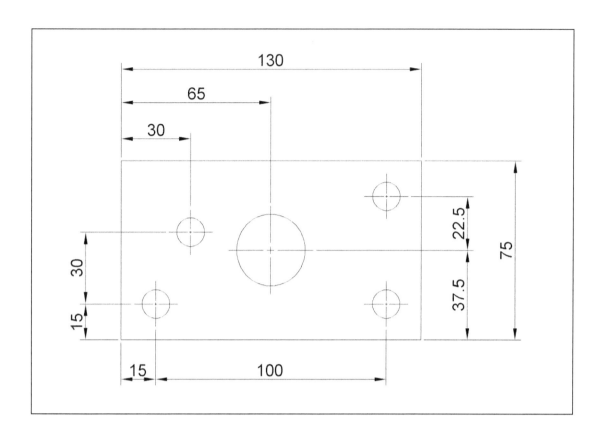

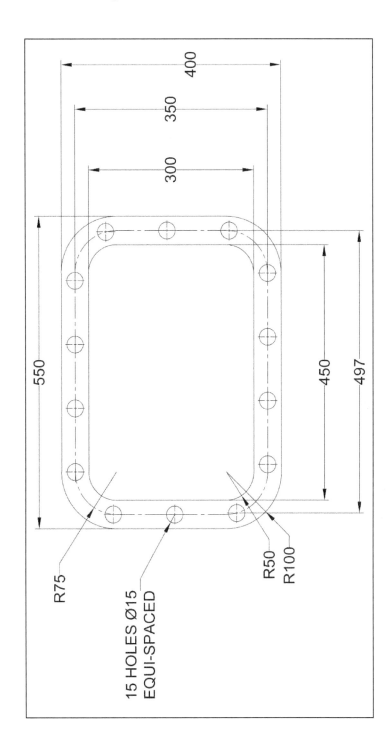

TurboCAD 2019 For Beginners

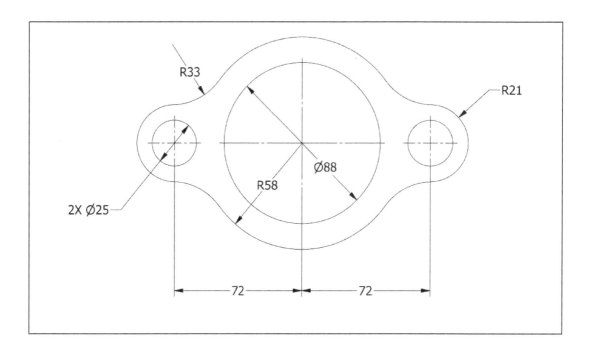

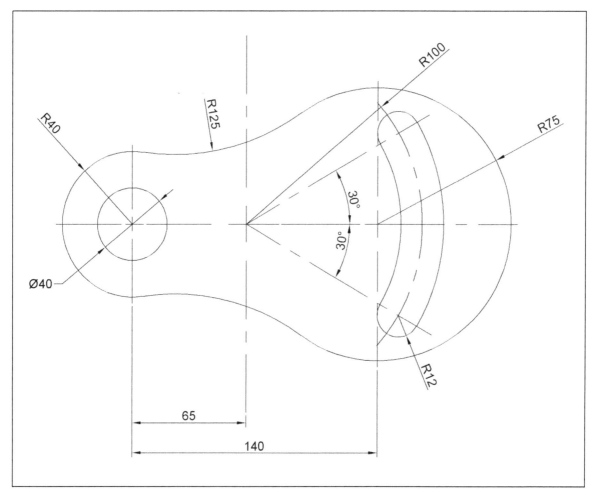

Editing Tools

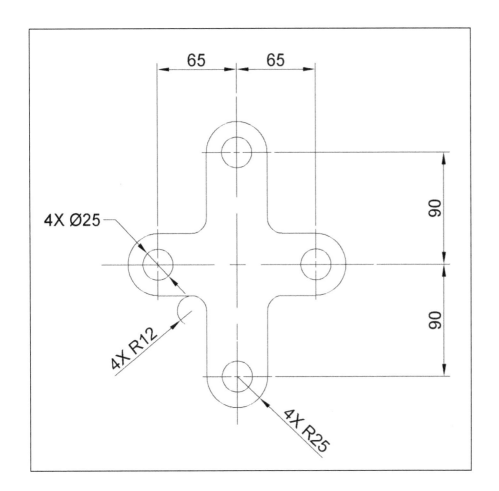

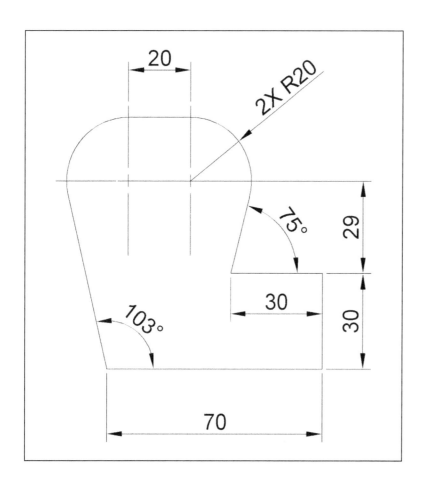

Chapter 5: Multi View Drawings

In this chapter, you will learn to create:

- **Orthographic Views**
- **Auxiliary Views**
- **Named Views**

Multi-view Drawings

To manufacture a component, you must create its engineering drawing. The engineering drawing consists of various views of the object, showing its true shape and size so that it can be dimensioned clearly. This can be achieved by creating the orthographic views of the object. In the first section of this chapter, you will learn to create orthographic views of an object. The second section introduces you to auxiliary views. The auxiliary views clearly describe the features of a component, which are located on an inclined plane or surface.

Creating Orthographic Views

Orthographic Views are standard representations of an object on a sheet. These views are created by projecting an object onto three different planes (top, front, and side planes). You can project an object by using two different methods: **First Angle Projection** and **Third Angle Projection**. The following figure shows the orthographic views that will be created when an object is projected using the **First Angle Projection** method.

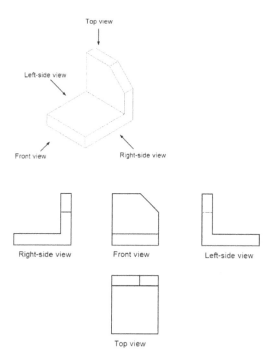

The following figure shows the orthographic views that will be created when an object is projected using the **Third Angle Projection** method.

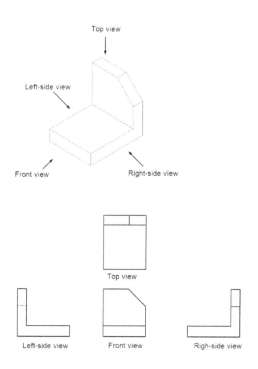

Example:

In this example, you will create the orthographic views of the part shown below. The views will be created by using the **Third Angle Projection** method.

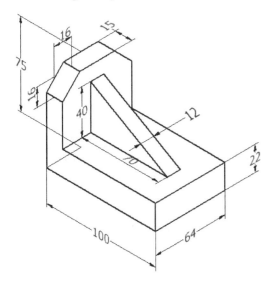

- Open a new drawing using the **ISO** template.
- Click **Format > Layers** on the menu bar; the **Layer** dialog appears.
- Click the **New Layer** button on the **Layer** dialog to create a new layer.
- Create new layers with the following properties.

Layer Name	Pen Width	Line Style
Object	0.20 mm	Continuous

- Click **Modes > Snaps > Ortho** on the menu bar.
- Click **View > Zoom > Zoom In** on the menu bar. Next, you need to draw construction lines. They are used as references to create actual drawings. You will create these construction lines on the **Construction** layer so that you can hide them when required.
- Click **Construction** drop-down > **Vertical Construction Line** on the **Draw** toolbar.

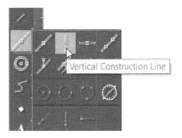

- Click in the graphics window on the left side, as shown.

- Click **Construction** drop-down > **Offset Construction Line** on the **Draw** toolbar.
- Click on the vertical line.
- Move the pointer toward the right.

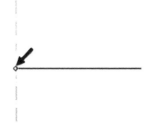

- Press the TAB key and type **100** in the **Offset** box.
- Press ENTER.

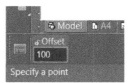

- Select the vertical construction line.
- Move the pointer towards the right.

- Click to create an offset line.

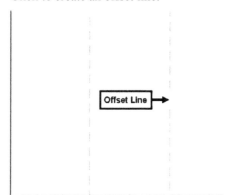

- Click the **Finish** icon on the Inspector Bar.

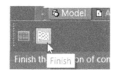

- Click **Construction Line > Horizontal Construction Line** on the **Draw** toolbar.
- Click on the bottom side of the graphics window, as shown.

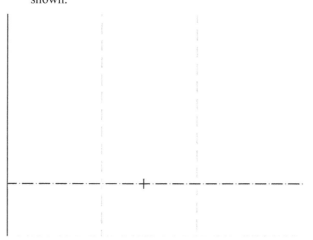

- Click **Construction** drop-down > **Offset Construction Line** on the **Draw** toolbar.
- Click on the horizontal construction line.
- Move the pointer toward upward.
- Press Tab key.
- Type 75 in the Offset box and press ENTER.
- Select the horizontal construction line.
- Move the pointer upward and click.

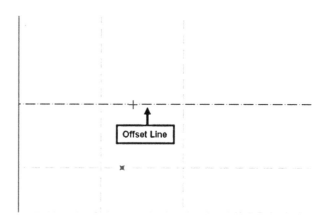

- Right-click and select **Finish**.
- Likewise, create other offset lines, as shown below. The offset dimensions are displayed in the image. Do not add dimensions to the lines.

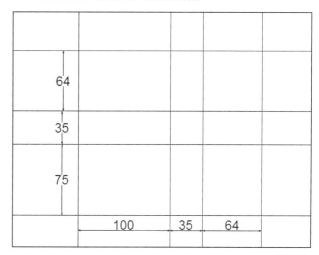

- Click the **Layers** drop-down > **Object** on the **Layer** toolbar.

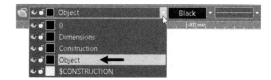

Now, you need to create object lines.

- Click **Line** on the **Draw** toolbar.
- Select the intersection points of the construction lines, as shown.

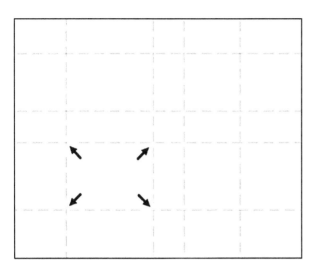

- Select the **Close** option from the command line to create the outline of the front view.

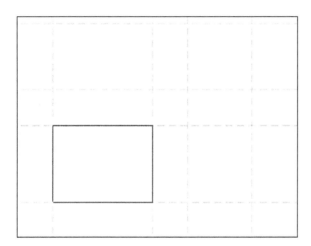

- Likewise, create the outlines of the top and side views.

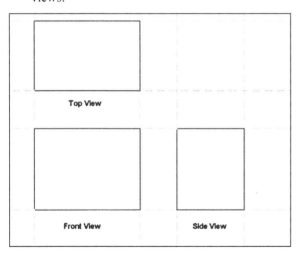

Next, you must turn off the **Construction** layer.

- Click **Format > Layer > Layers** on the menu bar; it opens **Layer** dialog.
- Click the **Visible** icon of the **$Construction** layer; the layer will be turned off.

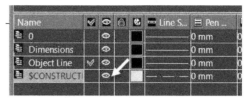

- Click **Line > Parallel** on the **Draw** toolbar.

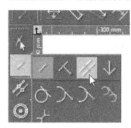

- Select the left vertical line of the **Front View**.
- Move the pointer towards the right and press Tab key.
- Type-in **15** in the **Offset** box and press ENTER.

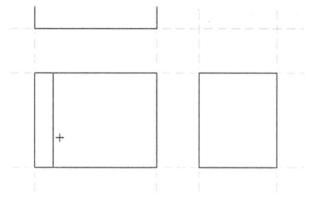

- Likewise, select the horizontal line of the **Front View**.
- Move the pointer above and press **Tab** key.
- Type-in **22** in the **Offset** box and press ENTER.

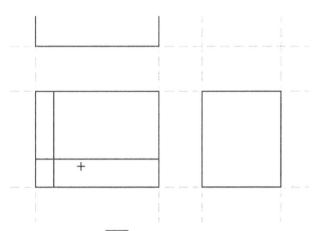

- Use the **Trim** tool and trim the unwanted lines of the front view, as shown below.

- Click **Modify 2D Objects** drop-down > **2D Fillet** on the **Modify** toolbar.
- Press the TAB key.
- Type 0 in the **Radius** box and press ENTER.
- Select the two lines forming a corner, as shown.

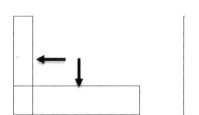

- Click **Line > Parallel** on the **Draw** toolbar.
- Select the small horizontal line of the front view, as shown.

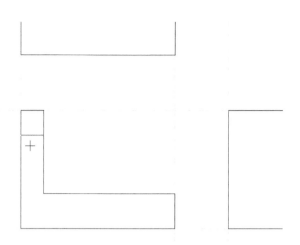

- Move the pointer towards the down and press **Tab** key.
- Type **-16** in the **Offset** box and press ENTER.

- Use the **Parallel** tool from the **Line** drop-down and create offset lines in the Top View, as shown below.

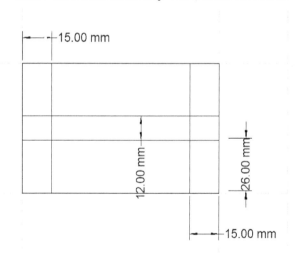

- Use the **Trim** tool and trim unwanted objects.

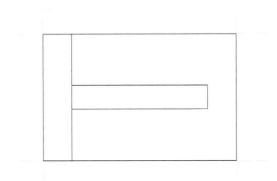

- Create other offset lines and trim the unwanted portions, as shown below.

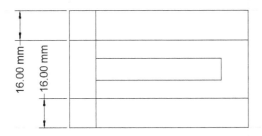

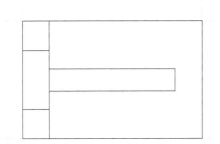

- Deactivate the **Ortho** icon on the **Modes** tab.
- Click **Line > Parallel** on the **Draw** toolbar.
- Select the horizontal line, as shown.
- Move the pointer upward and press the TAB key.
- Type 40 in the Offset box and press ENTER.

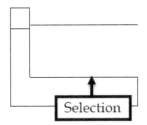

- Select the right vertical line of the front view, as shown.
- Select the endpoint on the top view, as shown.

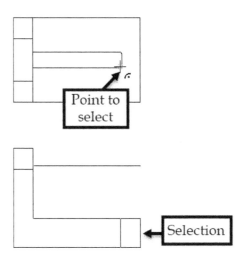

- Deactivate the **Ortho** icon on the **Snaps** toolbar.
- Click **Line** on the **Draw** toolbar.
- Select the endpoints of the horizontal and vertical lines, as shown.

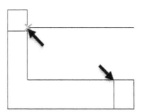

- Click the **Erase** icon on the **Workspace** toolbar,
- Select the vertical and horizontal lines, as shown.
- Press ENTER.

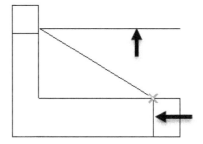

Next, you must create a right-side view. To do this, you must draw a 45- degree miter line and project the measurements of the top view onto the side view.

- Click the **Layers** drop-down on the **Layers** toolbar.
- Click the **Visible** icon next to the $CONSTRUCTION layer.

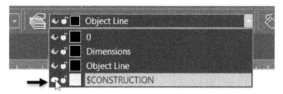

- Draw an inclined line by connecting the intersection points of the construction lines, as shown below.

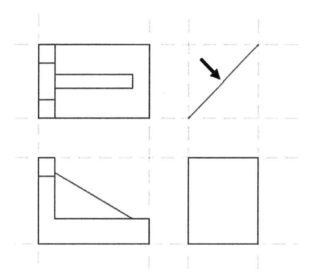

- Click **Construction > Horizontal Construction Line** on the **Draw** toolbar.
- Select the points on the top and front view, as shown below.

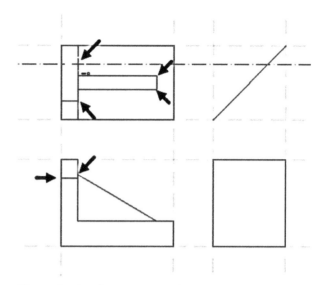

The projection lines are created, as shown below.

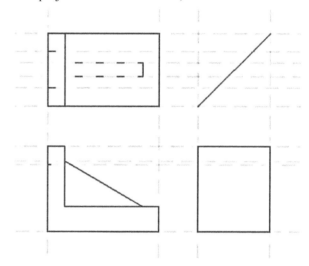

- Right-click and select **Finish** to exit the **Horizontal Line** tool.
- Click **Construction > Vertical Construction Line** on the **Draw** toolbar.
- Create vertical construction lines by selecting the intersection points, as shown.

TurboCAD 2019 For Beginners

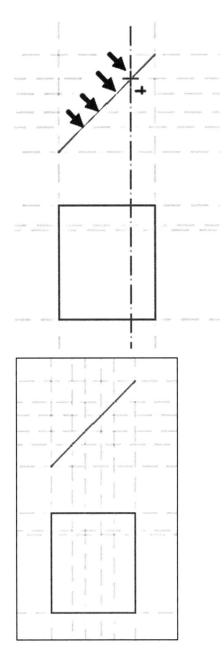

- Set the **Object** layer as current.
- Click **Line** drop-down > **Parallel** on the **Draw** toolbar.
- Select the lower horizontal line of the right view.
- Select Line from the local menu.

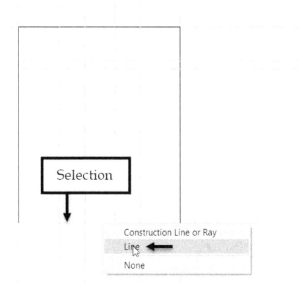

- Move the pointer upward and select the corner point of the front view, as shown.

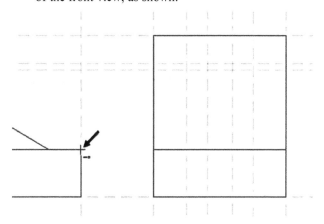

- Use the **Line** tool and create the objects in the side view, as shown below.

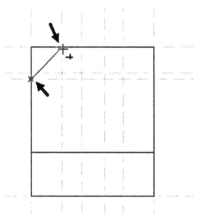

Multi View Drawings

TurboCAD 2019 For Beginners

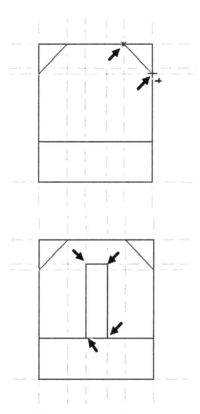

- Click the **Layer** drop-down on the **Layers** toolbar.
- Turn off the **$CONSTRUCTION** layer by clicking on the visible icon next to it.
- Trim the unwanted portions on the right-side view.

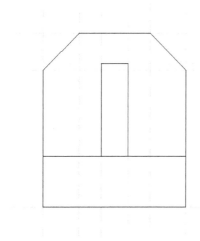

- Delete the inclined line, as shown.

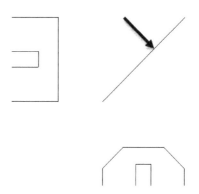

The drawing after creating all the views is shown below.

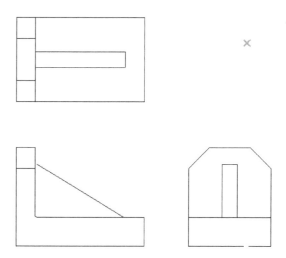

- Save the file as **ortho_views.tcw**. Close the file.

Creating Auxiliary Views

Most of the components are represented by using orthographic views (front, top, and side views). However, many components have features located on inclined faces. You cannot get the true shape and size for these features by using the orthographic views. To see an accurate size and shape of the inclined features, you must create an auxiliary view. An auxiliary view is created by projecting the component onto a plane other than horizontal, front, or side planes. The following figure shows a component with an inclined face. When you create orthographic views of the component, you will not be able to get the true shape of the hole on the inclined face.

TurboCAD 2019 For Beginners

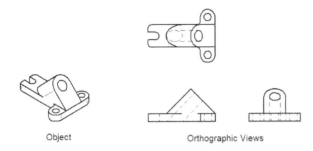

Object   Orthographic Views

To get the actual shape of the hole, you must create an auxiliary view of the object, as shown below.

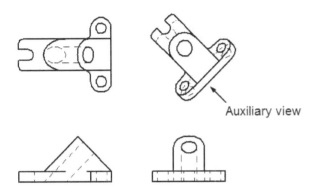

Auxiliary view

Example:

In this example, you will create an auxiliary view of the object shown below.

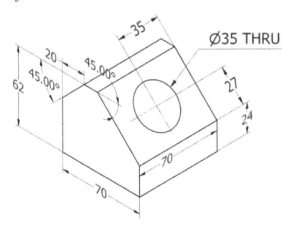

- Open a new TurboCAD file.
- Create four new layers with the following properties.

Layer Name	Lineweight	Linetype
Construction	0.00 mm	Continuous
Object	0.50 mm	Continuous
Hidden	0.20 mm	HIDDEN
Centerline	0.20 mm	CENTER

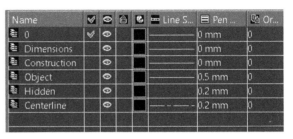

- Select the **Construction** layer from the **Layer** drop-down on the **Layers** toolbar.

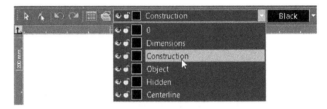

- Create a rectangle at the lower-left corner of the graphics window, as shown in the figure.

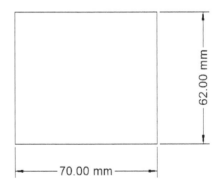

- Click the **Edit tool** on the **Draw** toolbar.
- Select the rectangle from the graphics window.
- Press the SPACEBAR.
- Click **Transform** drop-down > **Move** on the **Modify** toolbar.
- Select the lower-left corner of the rectangle as the base point.

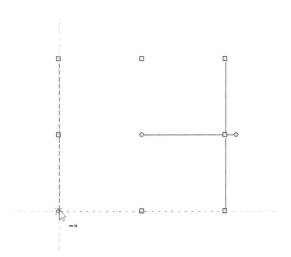

- Make sure that the **Ortho** is active on the **Modes** toolbar.
- Also, make sure that the **Keep Original Object** icon is selected.

- Move the pointer upward and press **Tab** key twice.

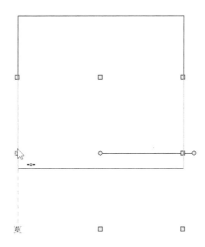

- Type **25** in the **Y Step** box on the Inspector Bar.

- Press **Enter** to copy the selected rectangle.

- Select the copied rectangle from the graphics window.

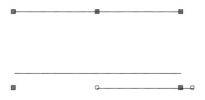

- Click **Transform** drop-down > **Rotate** on the **Modify** toolbar.
- Deselect the **Keep Original Object** icon on the Inspector Bar.
- Select the lower right corner of the copied rectangle as the base point.
- Move the pointer upward and select the top right corner of the rectangle.

TurboCAD 2019 For Beginners

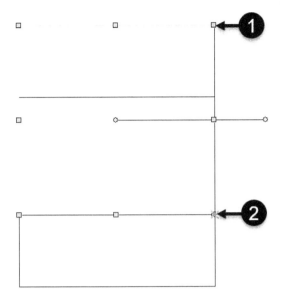

- Move the pointer toward left.

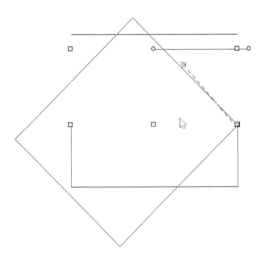

- Press **Tab** key and type-in **45** in the **Angle** box on the Inspector Bar.

- Press **Enter** to rotate the selected rectangle.
- Press **Esc** to deactivate the tool and click anywhere in the graphics window.

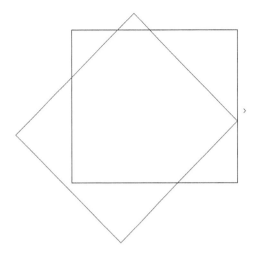

- Activate the **Rectangle** tool.
- Right-click and select **Local Snap > Extended Ortho**.

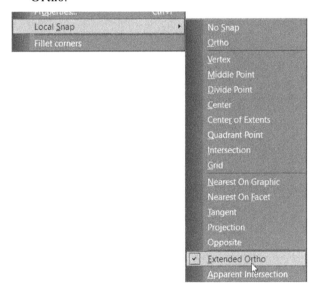

- Place the pointer on the top left corner of the existing rectangle.

84 | Multi View Drawings

TurboCAD 2019 For Beginners

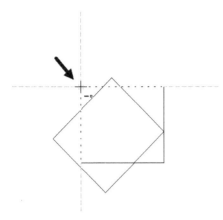

- Press Enter to position the rectangle.

- Move the pointer vertically upward, and then notice a vertical tracking line from the top left corner of the rectangle.
- Move the pointer along the tracking line up to an approximate distance.
- Click to specify the first corner of the rectangle.

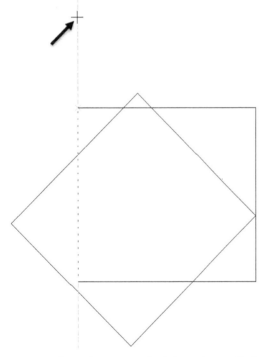

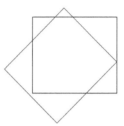

The rectangle located at the top is considered as top view and the below one as the front view.

- Click the **Edit tool** icon on the **Draw** toolbar.
- Select the newly created rectangle.
- Press the SPACEBAR.
- Click the **Explode** button on the **Modify** toolbar (or) click **Modify > Explode** on the menu bar.
- Click **Draw > Draw 2D > Line > Parallel** on the ribbon.
- Select the left vertical line of the top rectangle.
- Select any one of the through points, as shown; the selected vertical line is offset through the selected point.
- Again, select the left vertical line.
- Move the pointer, and then select the remaining through point.

- Move the pointer towards right and press **Tab** key on the keyboard.
- Type **70** in the **Size A** box on the Inspector Bar.
- Again, press **Tab** key and type **70** in the **Size B** box on the Inspector Bar.

85 | **Multi View Drawings**

TurboCAD 2019 For Beginners

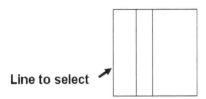

Line to select

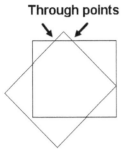

Through points

- Select the **Object** layer from the **Layer** drop-down in the **Layers** toolbar.
- On the **Property** toolbar, select **By Layer** from the **Line Width** drop-down.

- Deactivate the **Ortho** icon on the **Modes** toolbar.
- Activate the **Line** tool and select the intersection points on the front view, as shown.

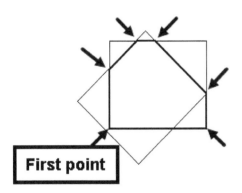

First point

- Likewise, create the object lines in the top view, as shown below.

- Select the **Construction** layer from the **Layers** toolbar.
- Click **Construction Line** > **Angular Construction Line** on the **Draw** toolbar.

- Click on the intersection points, as shown below.

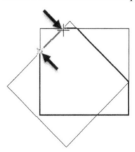

- Select the intersection point, as shown below.

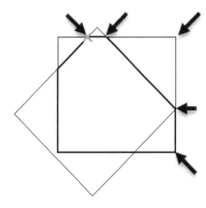

- Press ESC.

86 | Multi View Drawings

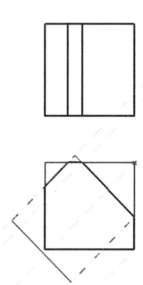
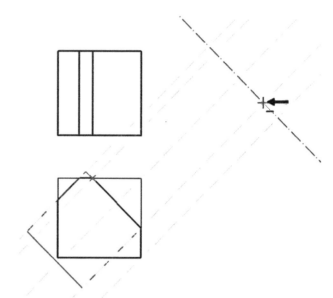

- Activate the **Angular Construction Line** tool.
- Select the intersection points, as shown.

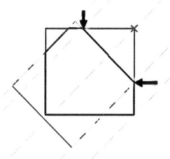

- Move the mouse pointer towards the right and click.
- Click the **Finish** icon on the Inspector Bar or Press **Enter** to create the construction line.

- Click **Construction Line** > **Parallel Construction Line** on the **Draw** toolbar.
- Click the **Through** icon on the Inspector Bar.

- Select the construction line created in the last step.

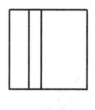

- Move the pointer toward the right.
- Press the TAB key and enter 35 in the **Offset** box.
- Press ENTER.

TurboCAD 2019 For Beginners

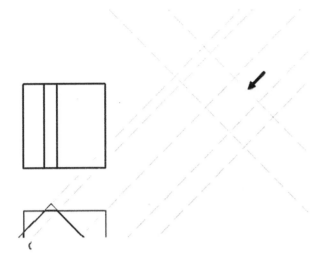

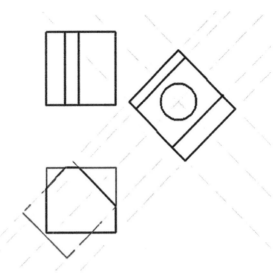

- Likewise, create another construction line at an offset distance of 35.

- Set the **Construction** layer as the current layer.
- Create projection lines from the circle.

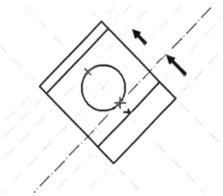

- Click the **Finish** icon on the Inspector Bar.
- Set the **Object** layer as the current layer. Next, create the object lines using the intersection points between the construction lines.

- Click the **Line** icon on the **Draw** toolbar.
- Set the **Hidden** layer as the current layer.

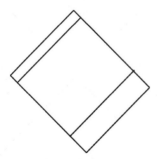

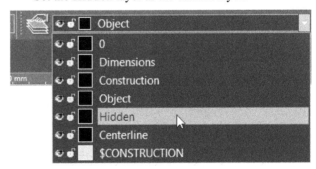

- On the **Property** toolbar, select the **By Layer** option from the **Line Pattern** drop-down.

- Use the **Circle** tool and create a circle of 35 mm in diameter.

88 | Multi View Drawings

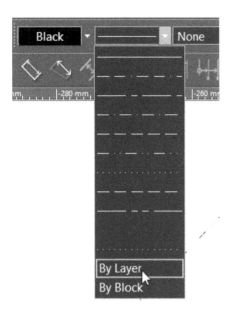

- Create the hidden lines, as shown.

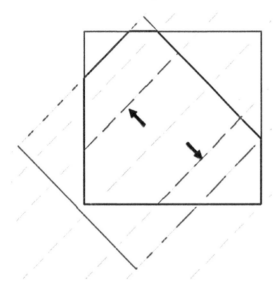

- Activate the **Line** tool.
- Set the **Centerline** layer as the current layer.
- On the **Property** toolbar, select the **By Layer** option from the **Line Pattern** drop-down.
- Select the **By Layer** option from the **Line Width** drop-down.

- Create the centerlines, as shown.

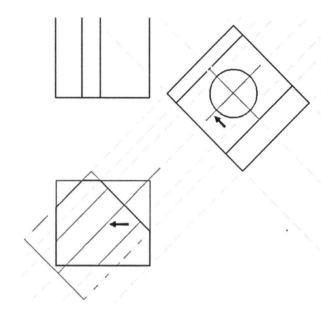

- Click **Construction** drop-down > **Vertical Ray** on the **Draw** toolbar.

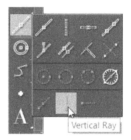

- Select the intersection point of the centerline and object line, as shown.

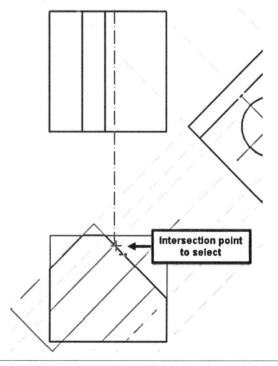

- Likewise, create two more rays, as shown.

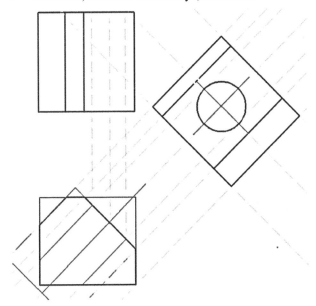

- Click **Construction** drop-down > **Horizontal Ray** on the **Draw** toolbar.

- Right click and select **Local Snap** > **Middle Point**.
- Select the middle point of the left vertical line of the top view.

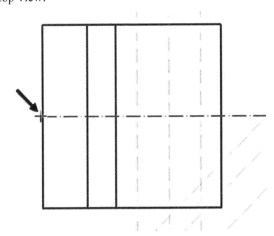

- Set the **Object** layer as the current layer.

- On the **Property** toolbar, select the **By Layer** option from the **Line Pattern** drop-down.
- Select the **By Layer** option from the **Line Width** drop-down.
- Click **Circle/Ellipse** > **Rotated Ellipse** on the **Draw** toolbar.

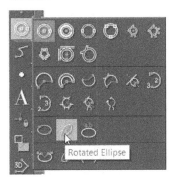

- Specify the first and second points, as shown.

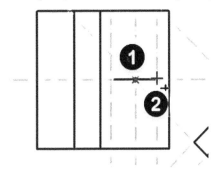

- Move the pointer downward and press **Tab** key thrice on the keyboard.
- Type **17.5** in the **Minor** box and press **Enter**.

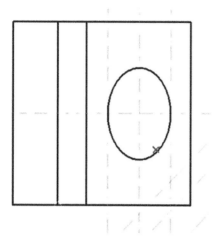

- Click the **Line** tool on the **Draw** toolbar.
- Set the **Centerline** layer as a current layer,
- Create the remaining centerlines.
- The drawing after hiding the **Construction** and **$CONSTRUCTION** layers is shown next.

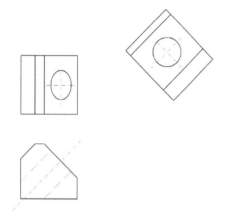

- Save the file as auxiliary_views.dwg.

Creating Named views

While working with a drawing, you may need to perform numerous zoom and pan operations to view key portions of a drawing. Instead of doing this, you can save these portions with a name. Then, restore the named view and start working on them.

- Open the **ortho_views.dwg** file (The drawing file created in the Orthographic Views section of this chapter).
- On the menu bar, click **View > Toolbars**.
- On the **Customize** dialog, click the **Toolbars** tab and check the **View** option.

- Click the **Zoom Window** icon on the **Zoom** toolbar.

- Create a window across the front view of the drawing.

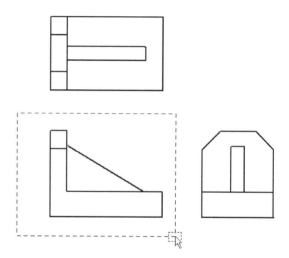

- Click the **Named View** icon on the **View** toolbar.

The **View** dialog appears.

TurboCAD 2019 For Beginners

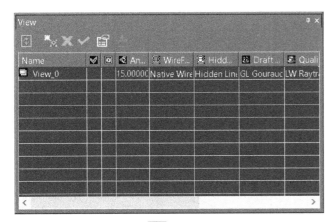

- Click the **New View** button on the **View** dialog; the **New View** dialog appears.
- Enter **Front** in the **New View** dialog

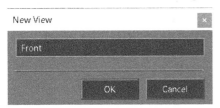

- Click **OK** on the **New View** dialog to accept.
- Likewise, create the named views for the top and right views of the drawing.

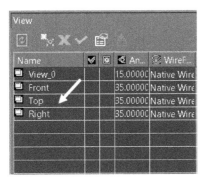

- To set the **Top** view to current, select it and click the **Activate** button on the **View** dialog.

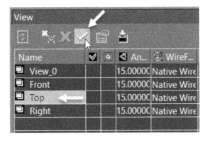

- Click the **Close** icon on the **View** dialog.

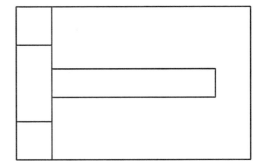

- Save and close the file.

Exercises

Exercise 1
Create the orthographic views of the object shown below.

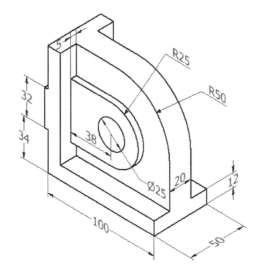

92 | Multi View Drawings

Exercise 2

Create the orthographic views of the object shown below.

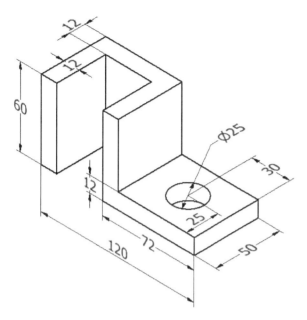

Exercise 4

Create the orthographic and auxiliary views of the object shown below.

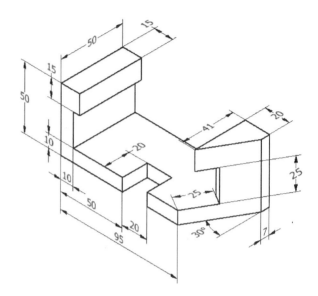

Exercise 3

Create the orthographic and auxiliary views of the object shown below.

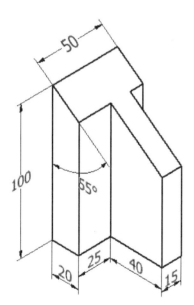

Chapter 6: Dimensions and Annotations

In this chapter, you will learn to do the following:

- **Create Dimensions**
- **Create Dimension Style**
- **Create Center Marks**
- **Add Dimensional Tolerances**
- **Add Geometric Tolerances**
- **Edit Dimensions**

Dimensioning

In previous chapters, you have learned to draw shapes of various objects and create drawings. However, while creating a drawing, you also need to provide the size information. You can provide the size information by adding dimensions to the drawings. In this chapter, you will learn how to create various types of dimensions. You will also learn about some standard ways and best practices of dimensioning.

Creating Dimensions

In TurboCAD, there are many tools available for creating dimensions. You can access these tools from the **Dimension** Toolbar and **Dimension** menu of the Menu bar.

TurboCAD 2019 For Beginners

The following table gives you the functions of various dimensioning tools.

Tool	Function
Smart 	This tool creates a dimension based on the selected geometry. • Create a rectangle, circle, arc, and two intersecting lines, as shown in the figure. 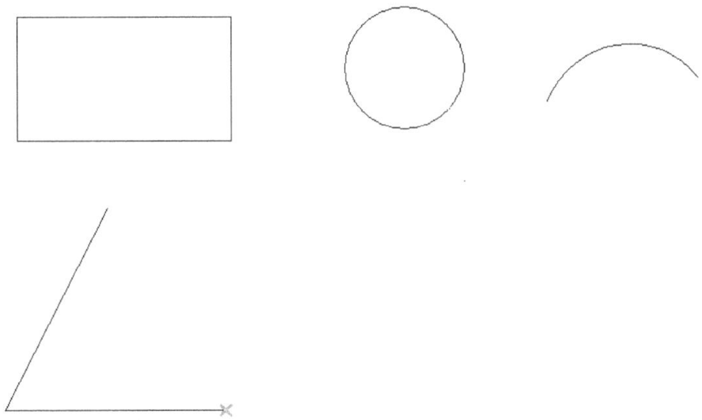 • Click **Dimension > Smart** on the Menu bar. • Select a line, move the pointer, and click to create the linear dimension. 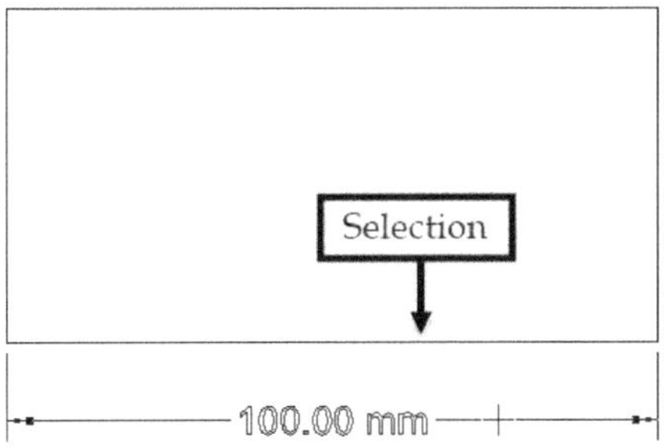 • Select a circle, move the pointer, and click to position the diameter dimension. 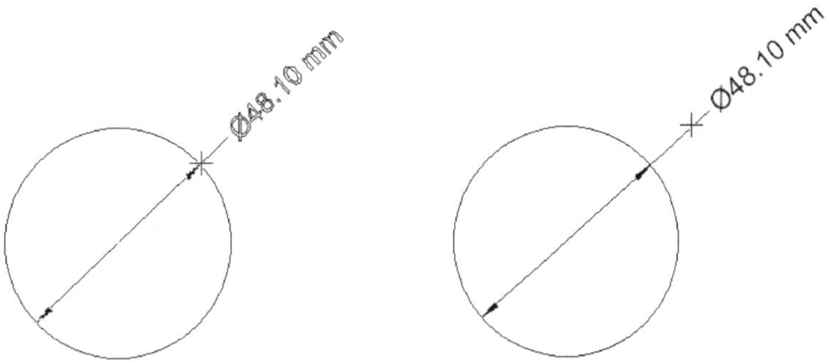 • Select an arc, move the pointer, and click to position the radial dimension.

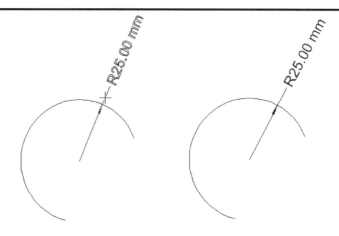

- Select two non-parallel lines and position the angular dimension between them.

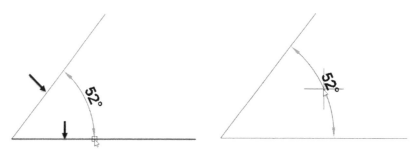

Likewise, you can create other types of dimensions using the **Smart** tool.

Orthogonal

This tool creates horizontal and vertical dimensions.

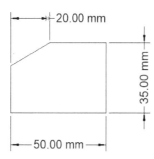

- Click the **Orthogonal** icon on the **Dimension** toolbar.
- Specify the first and second points of the dimension.
- Move the pointer in the horizontal direction to create a vertical dimension (or) move in the vertical direction to create a horizontal dimension.

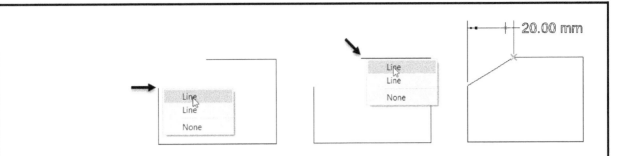

- Click to position the dimension.

(or)

- On the Inspector Bar, click the **Segment Dimensioning** icon.
- Select the line segment.
- Move the pointer in the vertical or horizontal dimension, and then click.

Parallel

This tool creates a linear dimension parallel to the object.

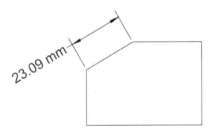

- Click the **Parallel** icon on the **Dimension** toolbar.
- Select the first and second points of the dimension line.

(or) Click the **Segment Dimensioning** icon on the Inspector Bar and then select the line segment to be dimensioned.

- Move the pointer and click to position the dimension.

Continuous

It creates a linear dimension from the second extension line of the previous dimension.

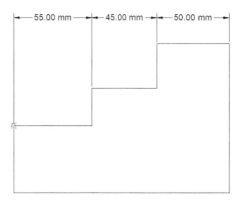

- Create a linear dimension by selecting the first and second points.

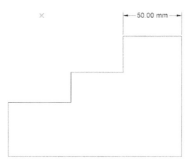

- Click **Continuous** on the **Dimension** toolbar.

- Click on the linear dimension at one of its ends; a chain dimension is attached to the pointer.

Dimensions and Annotations

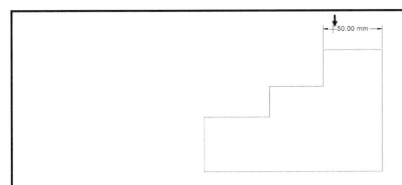

- Select the third and fourth points of the chain dimension.

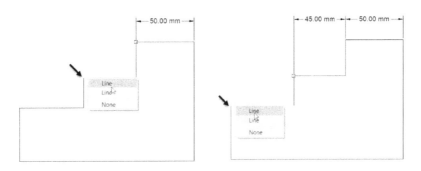

- Press Esc.

Baseline

It creates dimensions by using the previously created dimension, as shown below.

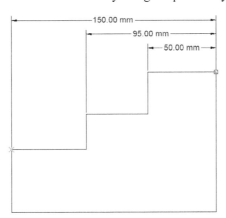

- Create a linear dimension.
- Click **Baseline** on the **Dimension** toolbar.

- Click on the linear dimension near its start point.

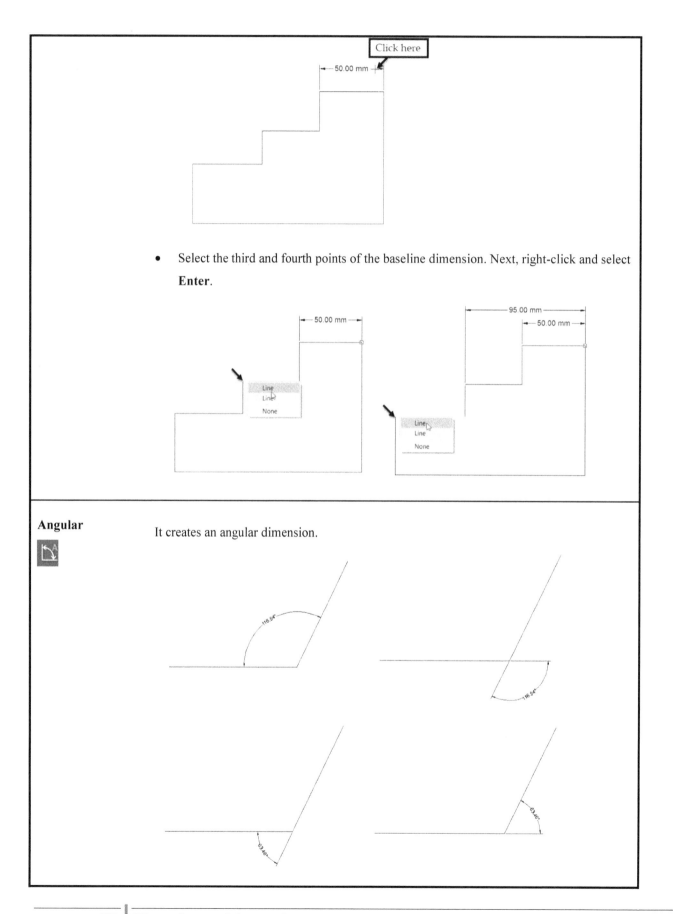

- Select the third and fourth points of the baseline dimension. Next, right-click and select **Enter**.

Angular   It creates an angular dimension.

- Click **Angular** on the **Draw** toolbar.
- Select the first and second lines.
- Move the pointer and position the angle dimension.
- To create an angle dimension on an arc, click on the arc near its start or endpoints. Next, move the pointer and click.

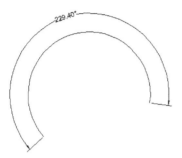

Diameter

It adds a diameter dimension to a circle or an arc.

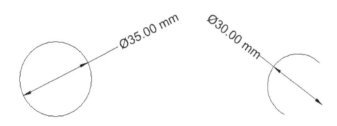

- Click **Diameter** on the **Dimension** toolbar.
- Select a circle or an arc and position the dimension.

Radius

It adds a radial dimension to an arc or circle.

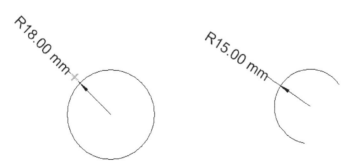

Large Radius

It creates jogged dimensions. A jogged dimension is created when it is not possible to show the center of an arc or circle.

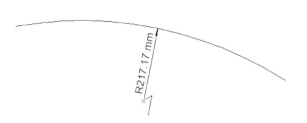

- Click **Radius** on the **Dimension** toolbar.
- Click the **Large Radius** icon on the Inspector Bar.

- Select an arc or circle.
- Specify the jog location.

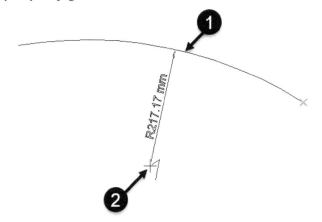

Center Mark

It adds a center mark to a circle or an arc.

- Click **Point** drop-down > **Center Mark** on the **Draw** toolbar.

- Select an arc or a circle; the center mark will be positioned at its center.

Centerline

It creates a centreline between two lines. The centreline has the associative property. It changes with the position of the lines

- Click **Point** > **Centerline** on the **Draw** toolbar.

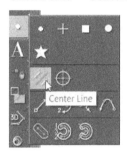

- Select two lines that are parallel or non-parallel to each other; a centreline is created between them.

- Change the position of the lines; the centreline also changes.

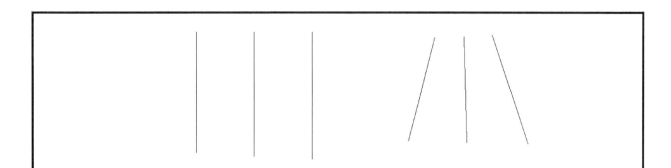

Datum

It creates ordinate dimensions based on the current position of the User Coordinate System (UCS).

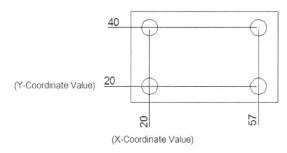

- Click **Datum** on the **Dimension** toolbar.
- Select the point of the object.
- Move the pointer in the vertical direction and click to position the X-Coordinate value.
- Select the point of the object.
- Move the pointer in the horizontal direction and click to position the Y-Coordinate value.

Quick

It dimensions one or more objects at the same time.

- Click **Dimension > Quick** on the Menu bar.
- Select one or more objects from a drawing.
- Right-click and select **FinishToSpecify**.

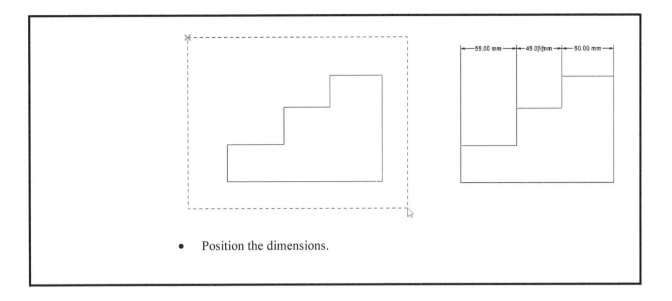

- Position the dimensions.

Example:

In this example, you will create the drawing, as shown in the figure, and add dimensions to it.

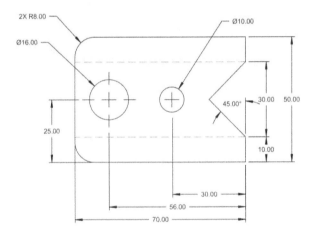

- Start a new drawing file using the ISO template.
- Create four new layers with the following settings.

Layer	Lineweight	Linetype
Object	0.50 mm	Continuous
Hidden	0.20 mm	HIDDEN
Dimensions	0.20 mm	Continuous

- Click **View > Zoom > Extents** on the menu bar.
- Create the drawing on the **Object** and **Hidden** layers.

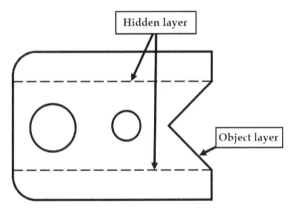

- Click the **Layers** icon on the **Layers** toolbar.
- Double-click on the **Dimensions** layer.

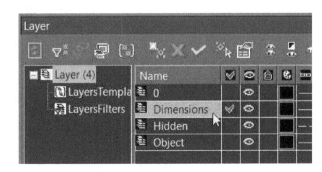

- Close the **Layers** palette.

Creating a Dimension Style

106 | Dimensions and Annotations

The appearance of the dimensions depends on the dimension style that you use. You can create a new dimension style using the **Style Manager** Palette. In this dialog, you can specify various settings related to the appearance and behavior of dimensions. The following example helps you to create a dimension style.

- On the Menu bar, click **Format > Style Manager Palette**.

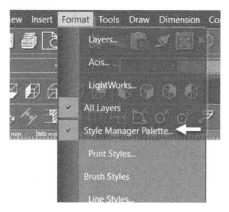

- On the **Style Manager** Palette, expand the **Dimension Styles** node and select the **Standard** option.
- Click the **Create New Style** icon on the **Style Manager** Palette.

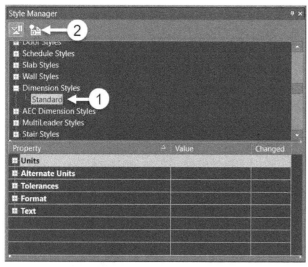

- On the **New Style Name** dialog, enter **Mechanical** in the **Enter unique style name** box.
- Click **OK**.

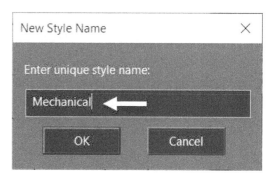

- In the **Property** section, expand the **Units** node.
- Ensure that the **Format for primary Dimension unit** is set to **Decimal**.
- Set **Precision for primary unit** to **0**.
- Type **.** in the **Decimal separator** box.

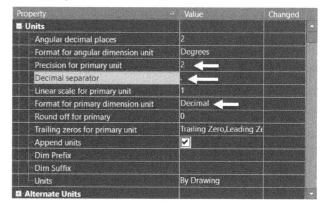

Study the other options in the **Units** node. Most of them are self-explanatory.

- Expand the **Text** node.
- Ensure that the **Text height** is set **3**.
- Set Text position horizontal to Centered.
- Set Text position vertical to **On line**.
- Check the **Force text horizontal inside align** option.

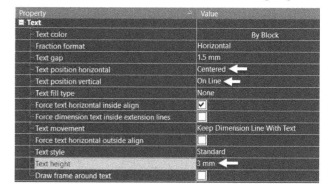

Study the other options in the **Text** node. These options let you change the appearance of the dimension text.

- Expand the **Format** node.
- Set the **Arrowheads size** to **3**.
- Set the **Center Mark Size** to **3**.
- In this node, notice the two options: **Extension line extension** and **Extension line offset**.

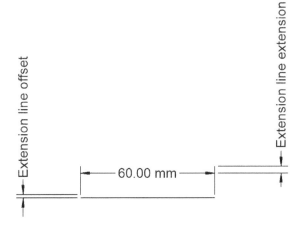

- Set **Extension line extension** and **Extension line offset** to **1.25**.
- Set the **Baseline increment** section to **5**.

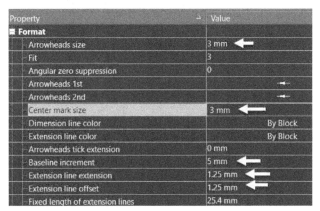

- Select **Center mark > Line**.
- Close the **Style Manager** Palette.
- On the Menu bar, click **Dimension > Diameter** (or) click the **Dimension** icon on the **Dimensions** toolbar.

- Select the large circle.
- Right-click and select **Properties** from the local menu.
- Click the **General** node on the **Properties** dialog.
- Select **Dimensions Styles > Mechanical**.

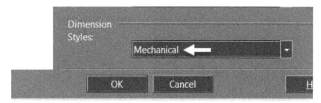

- Click the **Units/Tolerance** node.
- Uncheck the **Append Units** option.

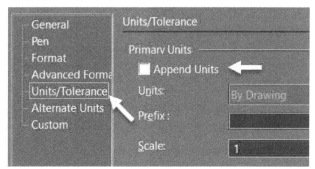

- Click **OK**.
- Move the pointer toward the top left corner and click.

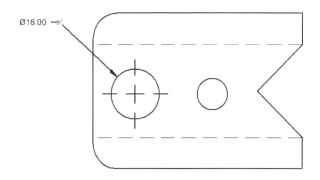

- Select the small circle.
- Move the pointer toward the top-right corner and click.

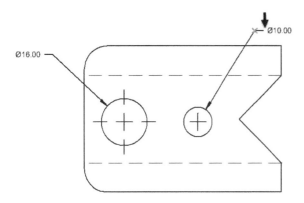

- Make sure that the **SNAP** option is turned on the status bar.

- Click the **Orthogonal** icon on the **Dimensions** toolbar.

- Select the lower right corner of the drawing.
- Select the **None** option from the menu.
- Select the quadrant point of the small circle; the dimension is attached to the pointer.
- Move the pointer vertically downwards and position the dimension, as shown below.

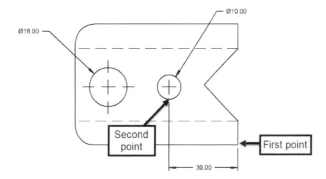

- Click the **Baseline** icon on the **Dimensions** toolbar.

- Select the right extension line of the linear dimension; a dimension is attached to the pointer.
- Select the endpoint of the center mark of the large circle; another dimension is attached to the pointer.
- Select the lower-left corner of the drawing.
- Press ENTER twice.

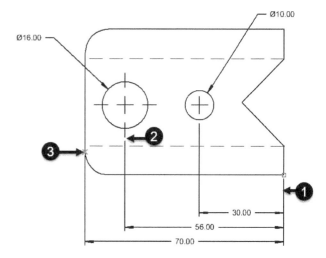

- Click the **Angular** icon on the **Dimensions** toolbar.

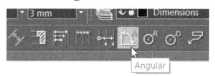

- Deactivate the **Vertex** icon on the **Snap Modes** toolbar.

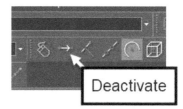

- Select the two lines of the drawing, as shown
- Position the angle dimension.

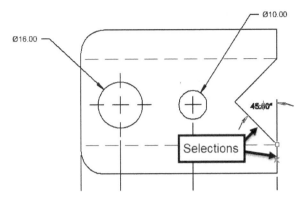

- Click the **Radius** icon on the **Dimensions** toolbar.

- Select the fillet located at the top left corner; the radial dimension is attached to the pointer.
- Next, position the radial dimension approximately at 45 degrees.

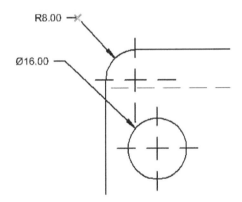

- Click the **Edit Tool** icon on the **Draw** toolbar.
- Select the radial dimension.
- Right-click and select **Properties**.
- On Properties dialog, type **2X** before the dimension **value** and press SPACEBAR.

- Select the **Advanced Format** node from the tree view.
- Select **None** from the **Center Marks** drop-down.

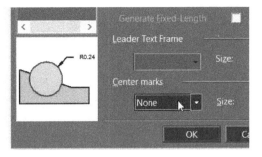

- Click **OK**.

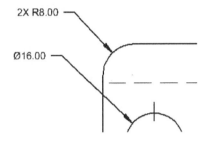

- Click in the graphics window to update the dimension text.
- Likewise, apply the other dimensions, as shown.
- Save and close the drawing.

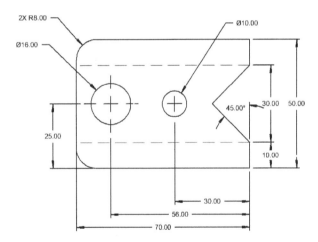

Adding Dimensional Tolerances

During the manufacturing process, the accuracy of a part is an important factor. However, it is impossible to manufacture a part with the exact dimensions. Therefore, while applying dimensions to a drawing, we provide some dimensional tolerances, which lie within acceptable limits. The following example shows you to add dimension tolerances in TurboCAD.

Example:

- Create the drawing, as shown below. Do not add dimensions to it.

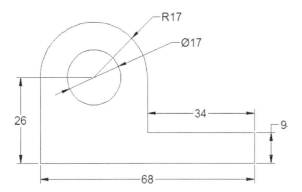

- Create a new dimension style with the name **Tolerances**.

- In the **Style Manager** palette, expand the **Tolerances** node.
- Set Tolerance **Precision** to 2.
- Check the **Append tolerance** option.
- Set the **Tolerance limit lower** and **Tolerance limit upper** to **0.05**.
- Set the **Tolerance pos ver** to **Middle**.

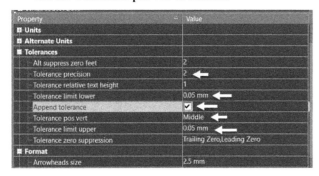

- Specify the following settings in the **Units**, **Text**, and **Format** nodes:

The **Units** node:
Format for primary dimension unit: Decimal
Precision for primary unit: 2
Decimal separator: .

The **Text** node:
Text Height: 2.5
Text position horizontal: Centered
Text position vertical: On line
Force text horizontal inside align: Checked
Append Units: Unchecked

The **Format** node:
Arrowheads size: 2.5
Center Mark: Line

- Close the **Style Manager** palette.
- On the Menu bar, click **Dimension > Diameter** (or) click the **Dimension** icon on the **Dimensions** toolbar.
- Right-click and select **Properties** from the local menu.
- Click the **General** node on the **Properties** dialog.
- Select **Dimensions Styles > Tolerances**.
- Click the **Units/Tolerance** node.
- Uncheck the **Append Units** option.
- Click the **Format** node and enter **2.5** in the **Height** box under the **Text** section.
- Click **OK**.
- Select the circle.
- Move the pointer toward the top left corner and click.

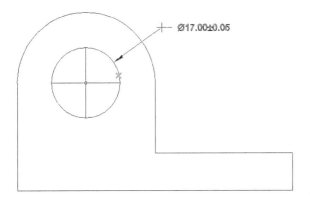

- Apply the remaining dimensions to the drawing.

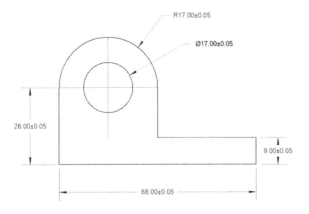

Geometric Dimensioning and Tolerancing

Earlier, you have learned how to apply tolerance to the size (dimensions) of a component. However, the dimensional tolerances are not sufficient for manufacturing a component. You must give tolerance values to its shape, orientation, and position as well. The following figure shows a note which is used to explain the tolerance value given to the shape of the object.

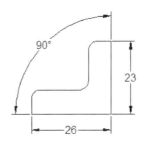

Providing a note in a drawing may be confusing. To avoid this, we use Geometric Dimensioning and Tolerance (GD&T) symbols to specify the tolerance values to shape, orientation, and position of a component. The following figure shows the same example represented by using the GD&T symbols. In this figure, the vertical face to which the tolerance frame is connected must be within two parallel planes 0.08 apart and perpendicular to the datum reference (horizontal plane).

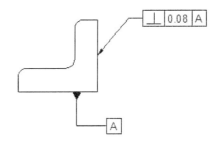

The Geometric Tolerance symbols that can be used to interpret the geometric conditions are given in the table below.

Purpose		Symbol
	Straightness	—

To represent the shape of a single feature.	Flatness	▱
	Cylindricity	⌭
	Circularity	○
	Profile of a surface	⌒
	Profile of a line	⌒
To represent the orientation of a feature with respect to another feature.	Parallelism	∥
	Perpendicularity	⊥
	Angularity	∠
To represent the position of a feature with respect to another feature.	Position	⌖
	Concentricity and coaxiality	◎
	Run-out	↗
	Total Run-out	↗↗
	Symmetry	⌯

Example 1: In this example, you will apply geometric tolerances to the drawing shown below.

113 | Dimensions and Annotations

TurboCAD 2019 For Beginners

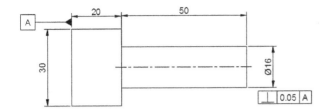

- Create the drawing, as shown below.

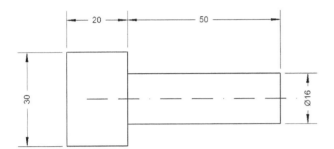

- Click **Tolerance** on the **Dimensions** toolbar.

- Right-click and select **Properties**; the **Properties** dialog appears.
- Select the **Tolerance** option from the tree.
- Click the upper box of the **Sym** group. The **Symbol** dialog appears.
- In the **Symbol** dialog, click the **Perpendicularity** symbol. The symbol appears in the **Sym** group.

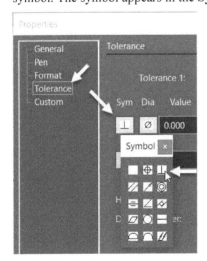

- Click in the **Value** box in the **Tolerance 1** group.

- Enter **0.05** in the box.

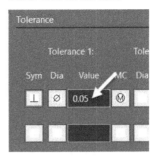

- Click in the **MC** box and select **None** from the **Material Condition** dialog.

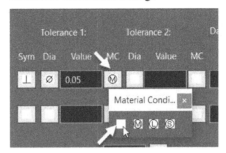

- Enter **A** in the upper box of the **Datum 1** group.

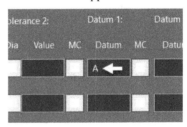

- Select the **Format** option from the tree.
- Type **2.5** in the **Text Height** box.
- Click **OK**.
- Click below the dimension, as shown.
- Press and hold the SHIFT key and move the pointer toward the right.
- Click to specify the second point, as shown.

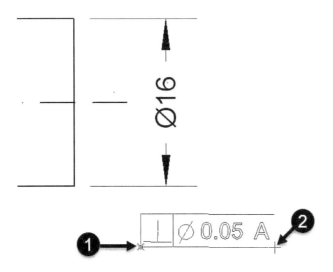
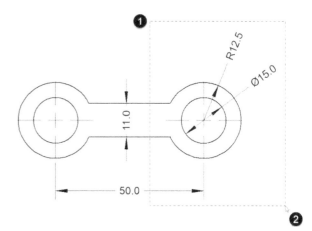

- Select the center point of the right-side circles.

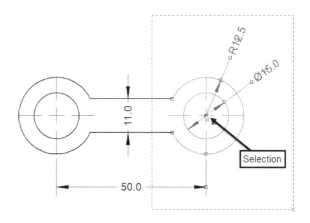

Editing Dimensions by Stretching

In TurboCAD, the dimensions are associated with the drawing. If you modify a drawing, the dimensions will be modified automatically. In the following example, you will stretch the drawing to modify the dimensions.

Example:

- Create the drawing, as shown below, and apply dimensions to it.

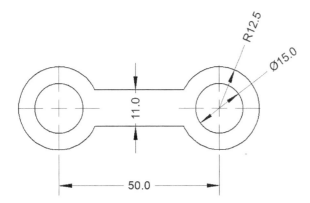

- Click **Modify > Stretch** on the menu bar.
- Drag a window and select the right-side circles and the horizontal lines.

- Move the pointer horizontally toward the right.
- Press the TAB key and type **30** in the **Length** box
- Press ENTER; the horizontal dimension is updated to 80.

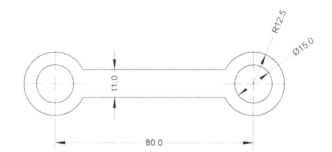

Modifying Dimensions using the Properties palette

Using the **Properties** dialog, you can modify the dimensional properties such as text, arrow size, precision, linetype, and lineweight. The **Properties** dialog comes in handy when you want to modify the properties of a particular dimension only.

Example:

- Create the drawing shown in the figure and apply dimensions to it.

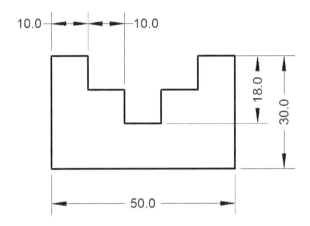

- Select the vertical dimension and right-click.
- Select **Properties** from the local menu; the **Properties** dialog appears.

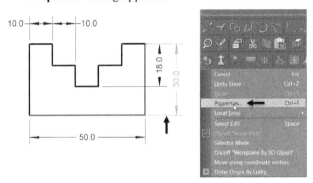

- In the **Properties** dialog, select the **Format** option.
- Under the **Arrowheads** section, set the **Size** to **2**.
- Enter 2 in the **Height** box.

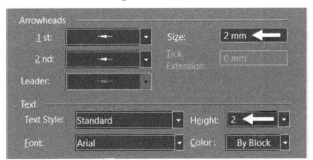

- Select the **Pen** option.
- Set **Color** to **Blue**.

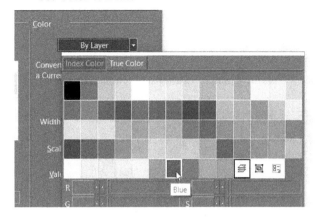

- Select the **Advanced Format** option.
- Under the **Extension Lines** section, set the **Extension** and **Offset** values to 2 and **1.25**, respectively.

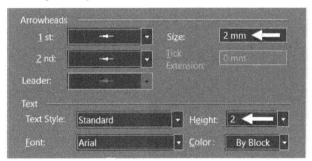

- Click **OK** on the **Properties** dialog; you will notice that the properties of the dimension are updated as per the changes made.

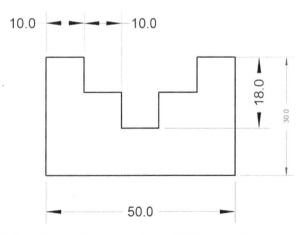

Matching Properties of Dimensions or Objects

In the previous section, you have learned to change the properties of a dimension. Now, you can apply these properties to other dimensions by using the **Format Painter** tool.

- Click the **Format Painter** icon on the **Draw** toolbar.

- Select the vertical dimension from the drawing.

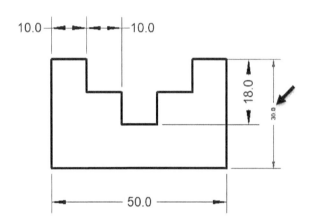

The **Format Painter** palette appears.

In this palette, you can select the settings that can be applied to the destination dimensions or objects. By default, all the options are selected in this palette.

- Select the other dimensions from the drawing; the properties of the source dimension are applied to other dimensions.

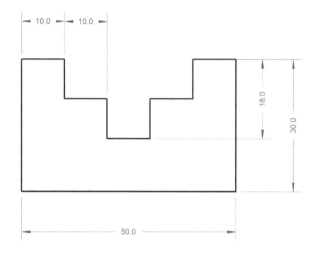

- Press ESC.

Exercises

Exercise 1

Create the drawing shown below and create hole callouts for different types of holes. Assume missing dimensions.

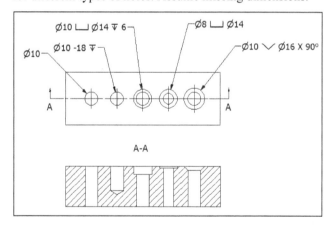

Exercise 2

Create the following drawings and apply dimensions and annotations. The Grid Spacing X= 10 and Grid Spacing Y=10.

TurboCAD 2019 For Beginners

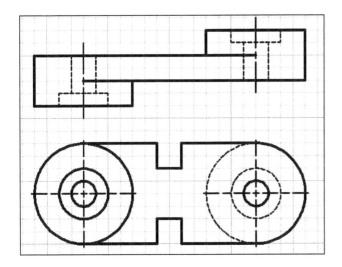

Method: Limits
Precision: 0.00
Upper Value: 0.05
Lower Value: 0.05

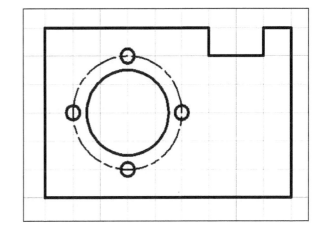

Exercise 3
Create the drawing shown below. The Grid spacing is 10 mm. After creating the drawing, apply dimensional tolerances to it. The tolerance specifications are given below.

Exercise 4
Create the drawing shown below.

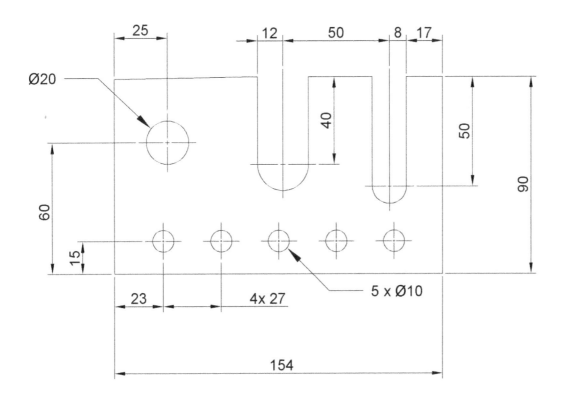

Exercise 5

Create the drawing shown below.

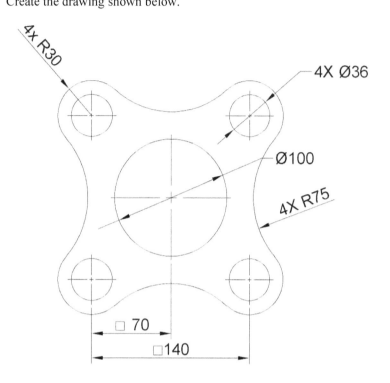

Chapter 7: Section Views

In this chapter, you will learn to:

- **Create Section Views**
- **Set Hatch Properties**

Section Views

In this chapter, you will learn to create section views. You can create section views to display the interior portion of a component that cannot be shown clearly using hidden lines. This can be done by cutting the component using an imaginary plane. In a section view, section lines, or cross-hatch lines are added to indicate the surfaces that are cut by the imaginary cutting plane. In TurboCAD, you can add these section lines or cross-hatch lines using the **Pick Point Hatching** tool.

The Pick Point Hatching tool

The **Pick Point Hatching** tool is used to generate hatch lines by clicking inside a closed area. When you click inside a closed area, a temporarily closed boundary will be created using the Polyline command. The closed boundary will be filled with hatch lines, and then it will be deleted.

Example 1:
In this example, you will apply hatch lines to the drawing, as shown in the figure below.

- Open a new TurboCAD file using the ISO template.
- Create four layers with the following properties.

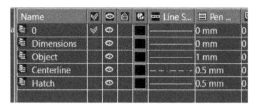

- Create the drawing, as shown below. Do not apply dimensions.

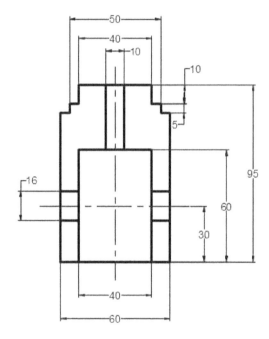

- Click the **Layers** icon on the **Layers** toolbar.
- Double-click on the **Hatch** layer.

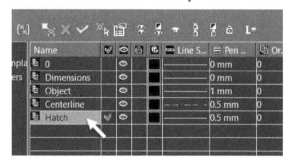

- Close the **Layer** palette.
- On the **Draw** toolbar, click **Hatch** drop-down > **Pick Point Hatching**.

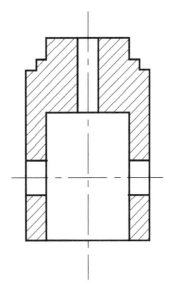

- Right-click and select **Properties**.
- On the **Properties** dialog, select the **Brush** option.
- Under the **Pattern** section, scroll down and select the **ANSI31** vector pattern.

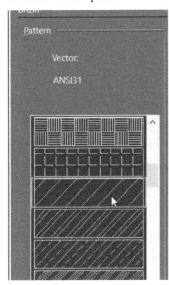

Example 2:

In this example, you will create the front and section views of a crank.

- Create five layers with the following settings:

Layer	Lineweight	Linetype
Object	1 mm	Continuous
Centerline	0.5 mm	CENTER
Hatch lines	0.5 mm	Continuous
Cutting Plane	1 mm	PHANTOM

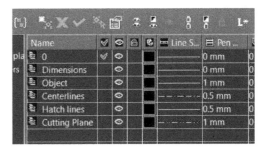

- Click on the four regions of the drawing, as shown below.

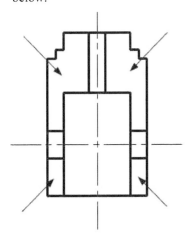

- Create construction lines using the **Construction line** and **Offset Construction line** commands, as shown.

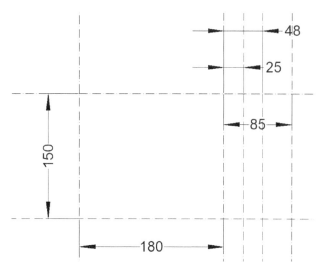

- Set the **Object** layer as current and create circles, as shown below.

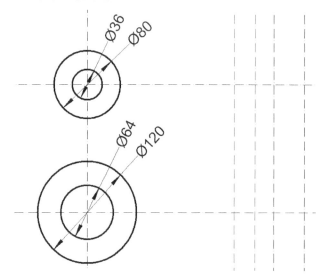

- Create construction lines, as shown.

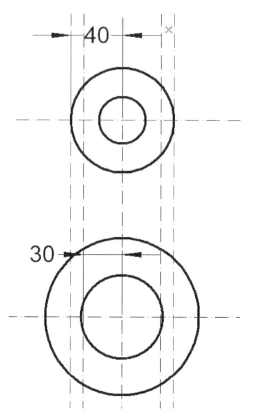

- Create two lines, as shown.

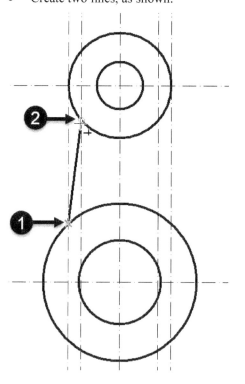

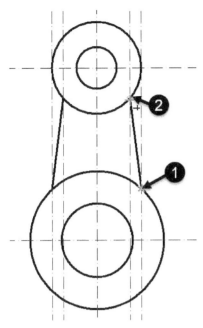

- On the **Modify** toolbar, click **Trim/ Extend** drop-down > **Fillet 2D**.
- Create the fillets of a 20 mm radius at the corners, as shown.

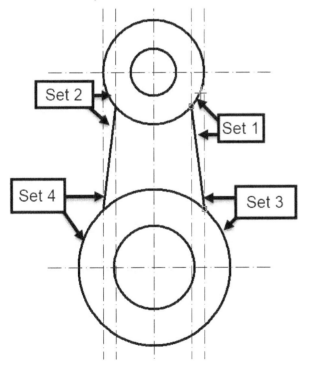

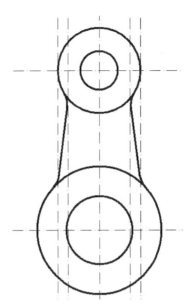

- On the **Draw** toolbar, click **Construction line** drop-down > **Offset Construction line**.
- Create three construction lines, as shown.

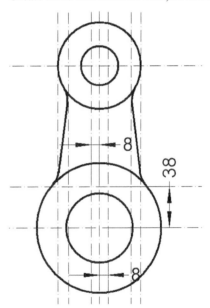

- On the **Draw** toolbar, click **Line**.
- Select the intersection points of the construction lines, as shown.

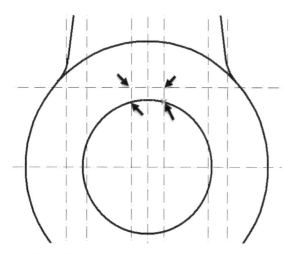

- Press Esc.

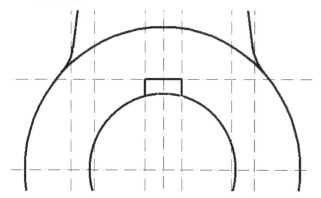

- On the **Modify** toolbar, click **Trim/ Extend** drop-down > **Trim**.

- Right-click and select **Select all**.
- Right-click and select **Finish Selection**.
- Trim the unwanted entities, as shown.

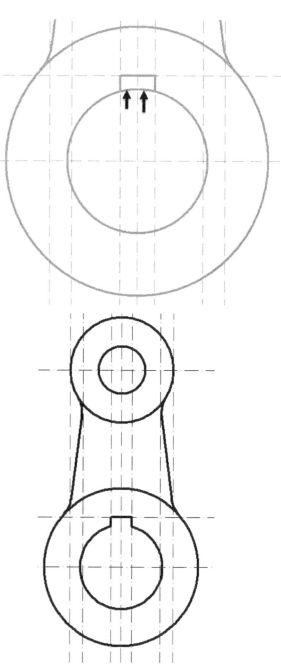

- On your own, create the objects of the section view, as shown below. (For any help, refer to the **Multi view Drawings** section of Chapter 5)

TurboCAD 2019 For Beginners

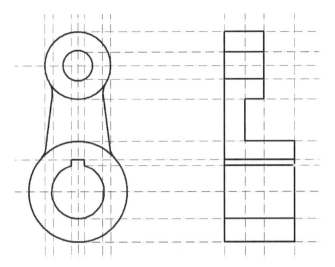

- Hide the **$CONSTRUCTION** layer.

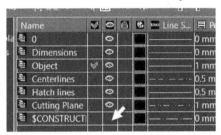

- On the **Draw** toolbar, click **Point** drop-down > **Center Line**.

- Set the **Centerlines** layer as current.
- On the **Property** toolbar, select **By Layer** from the **Line Pattern** and **Line Width** drop-downs.

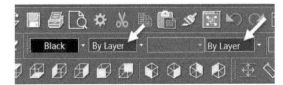

- Right-click and select **Properties**.
- On the **Properties** dialog, select the **Pen** option from the tree.

- Type **0.25** in the **Dash Scale** box.

- Click **OK**.
- Select the horizontal lines of the two holes, as shown.

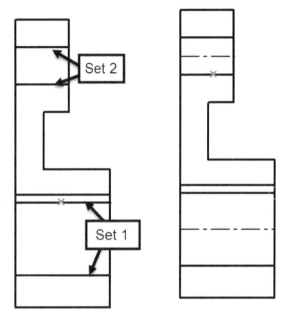

- On the **Draw** toolbar, click **Point** drop-down > **Center Mark**.

- Set the **Centerlines** layer as current.
- On the **Property** toolbar, select **By Layer** from the **Line Pattern** and **Line Width** drop-downs.

126 | Section Views

- Right-click and select **Properties**.
- On the **Properties** dialog, select the **Pen** option from the tree.
- Type **0.25** in the **Dash Scale** box.

- Click **OK**.
- Select the circles, as shown.

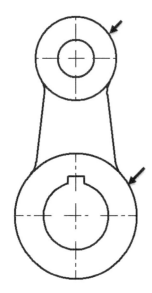

- On the **Draw** toolbar, click **Line** drop-down > **Polyline**.
- Set the **Cutting Plane** layer as current.
- On the **Property** toolbar, select **By Layer** from the **Line Pattern** and **Line Width** drop-downs.

- Right-click and select **Properties**.
- On the **Properties** dialog, select the **Pen** option from the tree.
- Type **0.5** in the **Dash Scale** box.
- Click **OK**.

- Activate the **Ortho** icon on the **Snap Modes** toolbar, if not already active.
- Pick a point below the front view, as shown.

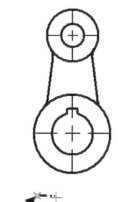

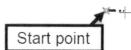

- Press the TAB key and type 20 in the **Length** box.
- Press the TAB key thrice.
- Type 0 in the **Start Width** box.
- Press the TAB key and type 8 in the **End Width** box.

- Press ENTER.
- Right-click and select **Local Snap > Extended Ortho**.
- Place the pointer on the lower quadrant point of the circle. The extension line appears from the quadrant point.

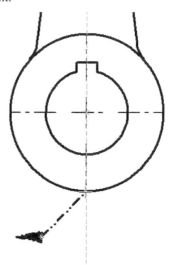

- Press and hold the Shift key, and then move the pointer toward the right.
- Click at the intersection point of the extension line.
- Move the pointer vertically up and click.

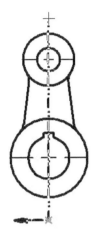

- Right-click and select **Local Snap > Extended Ortho**.
- Move the pointer to the endpoint of the lower horizontal line.

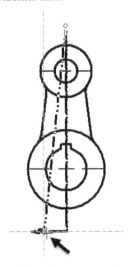

- Move the pointer upward.
- Move the pointer toward the left and click when extension lines are displayed from the endpoint of the lower horizontal line.

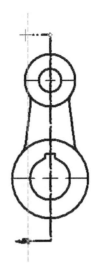

- Press the TAB key and enter 20 in the **Length** box.
- Press the TAB key and enter 180 in the **Angle** box.
- Press the TAB key twice and enter 8 in the Start Width box.
- Press the TAB key and enter 0.
- Press ENTER.

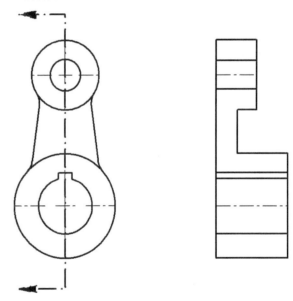

- Press Esc to deactivate the **Polyline** tool.
- On the **Draw** toolbar, click **Hatch** drop-down > **Pick Point Hatching**.

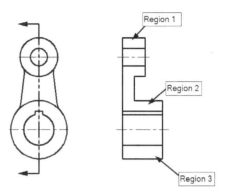

- Select **Hatch Lines** from the **Layers** drop-down available on the **Layers** toolbar.
- Select **By Layer** from the **Line Pattern** and **Line Width** drop-downs available on the **Property** toolbar.
- Right-click and select **Properties**.
- On the **Properties** dialog, select the **Brush** option.
- Under the **Pattern** section, scroll down and select the **ANSI31** vector pattern.

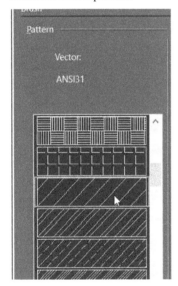

- Type **50** in the **Scale** box and click **OK**.
- Click in Region 1, Region 2, and Region 3.
- Press ESC.

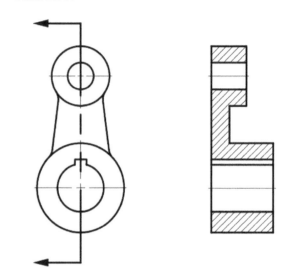

- Save the drawing as **Crank.dwg** and close.

The Hatch tool

While creating hatch lines, the island detection tools help you to detect the internal areas of a drawing.

Example:

- Create the drawing, as shown below. Do not apply dimensions.

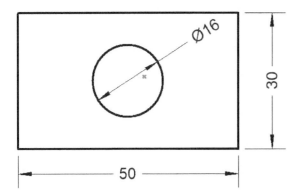

- On the **Draw** toolbar, click **Edit Tool**.
- Create a selection window across the entire drawing.
- Press the SPACEBAR.
- On the **Draw** toolbar, click **Hatch** drop-down > **Hatch**.

- Click on the hatch.
- On the **Property** toolbar, click on the **Brush Pattern** drop-down and select **ANSI31**.

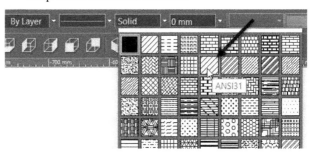

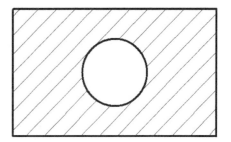

Exercises

Exercise 1
Create the half-section view of the object shown below.

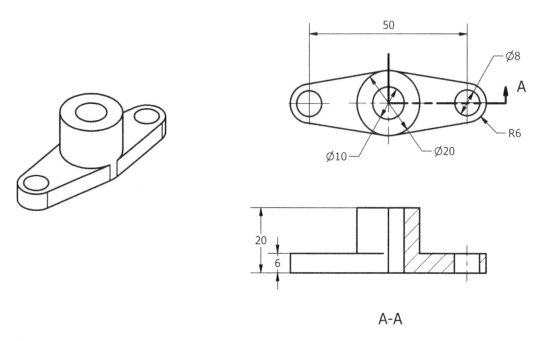

Exercise 2

In this exercise, the top, front, and right-side views of an object are given. Replace the front view with a section view. The section plane is given in the top view.

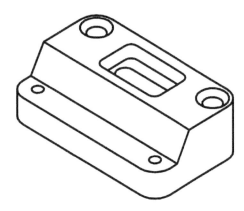

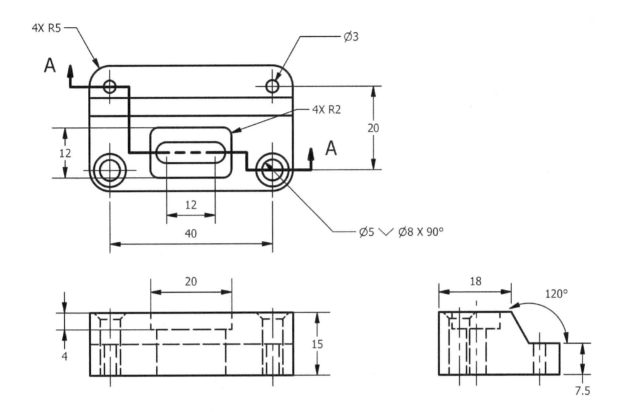

Chapter 8: Blocks, Attributes, and External References

In this chapter, you will learn to do the following:
- **Create and insert Blocks**
- **Use the Library Palette to insert Blocks**
- **Edit Blocks**
- **Define and insert Attributes**
- **Work with External References**

Introduction

In this chapter, you will learn to create and insert Blocks and Attributes in a drawing. You will also learn to attach external references to a drawing. The first part of this chapter deals with Blocks. A Block is a group of objects combined and saved together. You can later insert it in drawings. The second part of this chapter deals with Attributes. An Attribute is an intelligent text attached to a block. It can be any information related to the block, such as description, part name, and value. The third part of the chapter deals with external references. External references are drawing files, images, PDF files attached to a drawing.

Creating Blocks

To create a block, first, you need to create shapes using the drawing tools and use the Create Block command to convert all the objects into a single object. The following example shows the procedure to create a block.

Example 1
- Create the drawing, as shown below. Do not apply dimensions. Assume the missing dimensions.

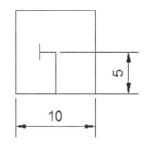

- On the **Draw** toolbar, click **Edit Tool**.
- Create a selection window across the entire drawing.
- Press the TAB key.
- On the **Draw** toolbar, click **Block** drop-down > **Create Block**.

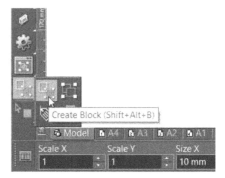

- Enter **Target** in the **Block Name** field.

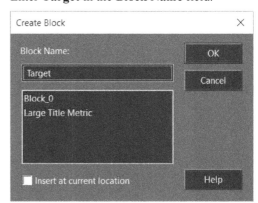

- Uncheck the **Insert at current location** option, if selected.
- Click **OK** on the **Create Block** dialog; the block will be created and saved in the database.

Inserting Blocks

After creating a block, you can insert it at the desired location inside the drawing using the **Blocks** palette. The procedures to insert blocks are explained in the following examples.

Example 1

- Click **Insert > Block** on the menu bar (or) click the **Blocks** icon on the **Draw** toolbar.

The **Blocks** palette appears.

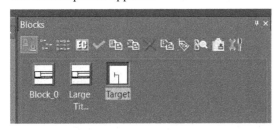

The BLOCKS palette can be used to access a large number of blocks.

- Click and drag the **Target** block from the **Blocks** palette into the drawing.

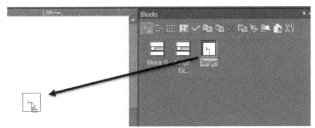

- Type 3 in the **Scale X** box and press TAB.
- Type 2 in the **Scale Y** box and press ENTER; the block will be scaled, as shown below.

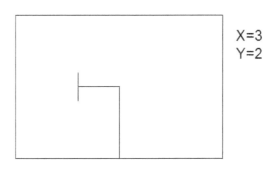

Example 2 (Rotating the block)

- Click and drag the **Target** block from the **Blocks** palette into the drawing.
- Click in the **Rot** box and type **45**.

- Press **ENTER**; the block will be rotated by **45** degrees.

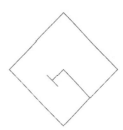

- Save and close the drawing file.

Replacing the Blocks

TurboCAD allows you to redefine an already-created block.

- Download the Replacing Blocks.tcw file from the companion website.
- Open the downloaded drawing file.

- On the **Draw** toolbar, click **Edit Tool**.
- Create a selection window across the objects located on the left side, as shown.

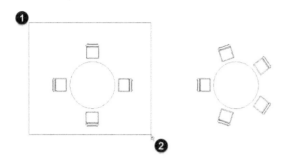

- Press the TAB key.
- On the **Draw** toolbar, click **Block** drop-down > **Create Block**.
- Type **Dining Table** in the **Block Name** box.
- Select the **Insert at current location** option.
- Click **OK** on the **Create Block** dialog; the block is created and inserted at its current location.

Now, you need to replace the Dining Table block with the objects located on the right side.

- Create a selection across the objects located on the right side.

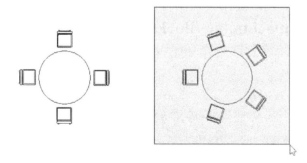

- Press TAB to accept the selection.
- On the **Draw** toolbar, click **Block** drop-down > **Create Block**.
- Type **Dining Table_A** in the **Block Name** box.
- Select the **Insert at current location** option.

- Click **OK** on the **Create Block** dialog; the block is created and inserted at its current location.
- In the **Blocks** palette, select the **Dining Table** block.
- Click the **Replace References** icon on the **Blocks** palette.

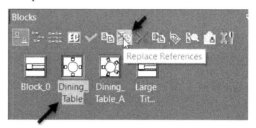

- On the **Replace Block Reference** dialog, select the **Dining_Table_A** block from the **Replace with** section.

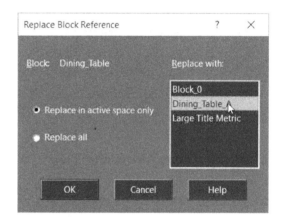

- Click **OK**. The first block is replaced with the second one.

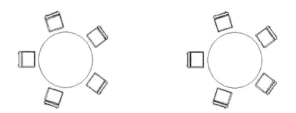

- Close the drawing file without saving.

Exploding Blocks

When you insert a block into a drawing, it will be considered as a single object, even though it consists of numerous individual objects. At many times, you may require to break a block into its individual parts. Use the **Explode** tool to break a block into its individual objects.

- Select the block to be exploded.
- Click **Modify > Explode** on the menu bar (or) click the **Explode** icon on the **Modify** toolbar.
  The block will be broken into individual objects. You can select the individual objects by clicking on them. (Refer to the Explode Tool section of Chapter 4).

Using the Library

Library palette is one of the additional means by which you can effectively insert blocks and drawings. Using the Library palette, you can insert blocks created in one drawing into another drawing. You can display the **Library** palette by using the **Customize** dialog. To do this, click **Tools > Customize** on the menu bar (or) click **Customize Controls** icon on the **Main** toolbar. On **Customize** dialog, click the **Palettes** tab and check the **Library** option from the **Palettes** list. Next, close the **Customize** dialog.

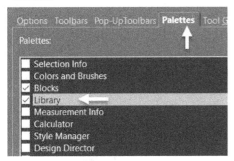

The following example shows you to insert blocks using the Library palette.

Example:

- Open a new drawing file.
- Create the following symbols and convert them into blocks. You can also download them from the companion website.

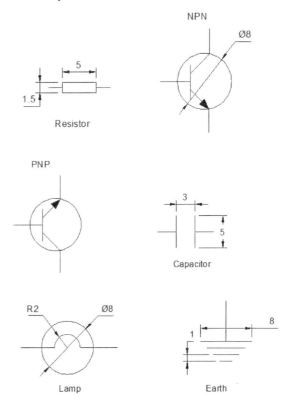

Variable Resistor

- Save the file as **Electronic Symbols.tcw**. Close the file.
- Open a new drawing file.
- In the **Library** palette, click the **Load Folder** icon.

- Browse to the location of the **Electronic Symbols.tcw** file using the **Look in** drop-down.
- Double-click on the **Electronic Symbols.tcw** file.

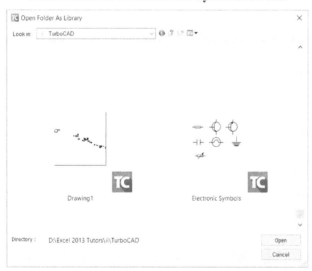

- Click **No** on the message box. The folder containing the **Electronic Symbols.tcw** file is added to the **Library** palette.
- On the **Library** palette, scroll to the newly added folder and double-click on it.
- Select the **Electronic Symbols.tcw** file from the **Library** palette.
- Click the **Blocks** tab at the bottom of the **Library** palette.

All the blocks present in the file will be displayed.

- Drag and place the blocks in the graphics window.

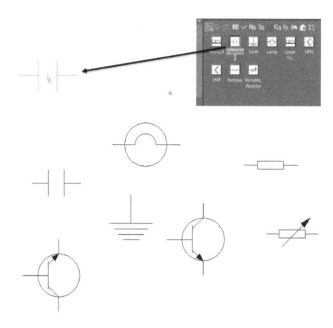

- Use the **Move** and **Rotate** tools and arrange the blocks, as shown below.

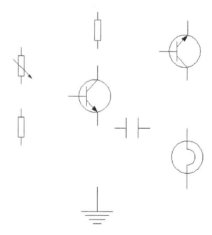

- Close the **Library** palette.
- Use the **Line** tool and complete the drawing, as shown below.

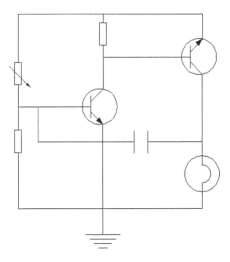

Editing Blocks

During the design process, you may need to edit blocks. You can easily edit a block using the **Block Editor** window. As you edit a block, all the instances of it will be automatically updated. The procedure to edit a block is discussed next.

- Create a block, as shown below.

- On the Menu bar, click **Insert** > **Blocks**.
- Click and drag the **Pump** block from the **Blocks** palette into the graphics window.
- Select the **Pump** block from the **Blocks** palette.
- Click the **EC** icon on the **Blocks** palette.

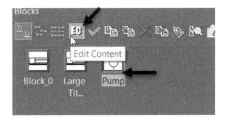

- Click **Draw** toolbar > **Line** drop-down > **Polyline** on the menu bar and draw a polyline, as shown below.

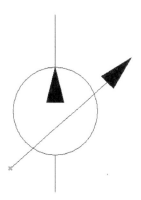

- Click **Finish Edit Content** on the **Blocks** palette.

Defining Block Attributes

A block attribute is a line of text attached to a block. It may contain any type of information related to a block. For example, the following image shows a Compressor symbol with an equipment tag. The procedure to create an attribute is discussed in the following example.

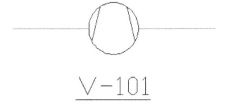

Example 1:
- Open a new drawing file.
- Create the symbols, as shown below.

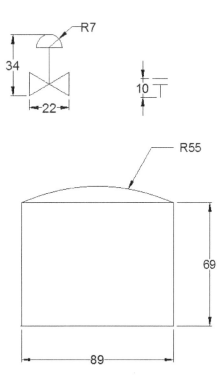

- Create three blocks with names **Control Valve**, **Nozzle**, and **Tank**.

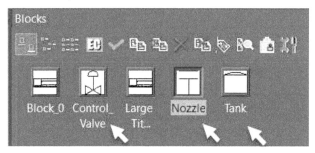

- Select the **Control Valve** block from the **Blocks** palette.
- Click the **EC** icon.

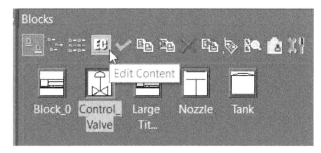

- On the **Draw** toolbar, click **Blocks** drop-down > **Block Attribute**.

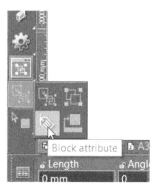

- Specify the location of the attribute below the control valve symbol.
- Type VALVETAG.

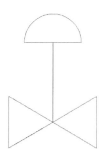

- Click in the **Height** box and type **6**.
- Click in the **Prompt** box and type **Specify valve tag**.
- Click in the **Default** box and type **CV-001**.

- Click the **Finish** icon.
- Click the **Finish Edit Content** icon on the **Blocks** palette.
- Select the **Tank** block from the **Blocks** palette.
- Click the **EC** icon on the **Blocks** palette.
- On the **Draw** toolbar, click **Blocks** drop-down > **Block Attribute**.

- Specify the location of the attribute inside the Tank symbol.
- Type EQUIPMENTAG.

- Click in the **Height** box and type **6**.
- Click in the **Prompt** box and type **Specify equipment tag**.
- Click in the **Default** box and type **TK-001**.

- Click the **Finish** icon.
- Click the **Finish Edit Content** icon on the **Blocks** palette.

Inserting Attributed Blocks

You can use the Blocks palette to insert the attributed blocks into a drawing. The procedure to insert attributed blocks is discussed next.

- On the Menu bar, click **Insert > Blocks**.
- Click and drag the **Control Valve** block from the **Blocks** palette into the graphics window.
- Leave the default value in the **Attribute Value** box, and then click **Close**.

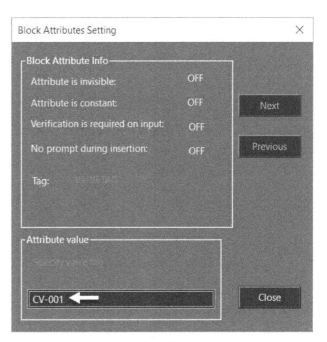

- Click and drag the **Tank** block from the **Blocks** palette into the graphics window.

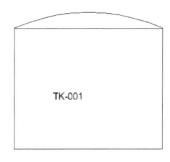

- Leave the default value in the **Attribute Value** box, and then click **Close**.
- Click and drag the **Control Valve** block from the **Blocks** palette into the graphics window.

TurboCAD 2019 For Beginners

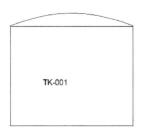

- Type **CV-002** in the **Attribute Value** box, and then click **Close**.
- Select the **Tank** block from the **Blocks** palette.
- Click the **EC** icon on the **Blocks** palette.
- Click and drag the **Nozzle** block from the **Blocks** palette into the graphics window.
- Rotate and move the nozzle to the location, as shown.

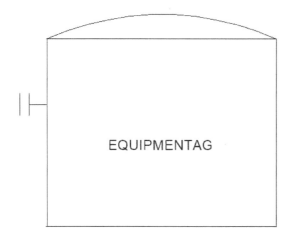

- Likewise, insert another Nozzle block and position it, as shown.

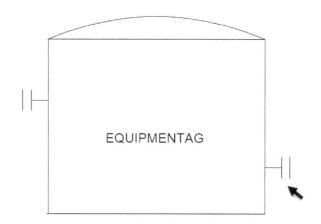

- Click the **Finish Edit Content** icon on the **Blocks** palette.
- Use the **Polyline** tool and connect the control valves and tank.

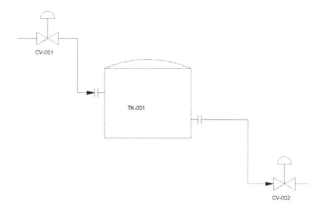

Working with External references

In TurboCAD, you can attach a drawing file, image, or pdf file to another drawing. These attachments are called External References. They are dynamic and update automatically when changes are made to them. In the following example, you will learn to attach drawing files to a drawing.

Example 1:
- Create the drawing shown below.

141 | Blocks, Attributes, and Xrefs

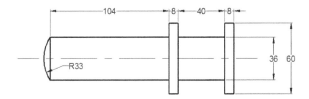

- Save the drawing as **Crank pin.tcw**
- Create another drawing as shown below (For help, refer to the **Multi View Drawings** section in Chapter 5).

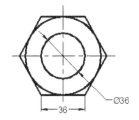

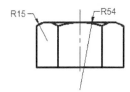

- Save the drawing as **Nut.tcw** and close it.
- Open the **Crank.tcw** file created in Chapter 8.
- Click **Insert > Create External Reference** on the menu bar; the **External Reference Files** dialog appears.
- Browse to the location of the **Crank pin.tcw** and double-click on it.
- On the **Create External Reference** dialog, select **Type > Attach**.
- Select **Center of Xref Extents** from the **Reference Point** section.

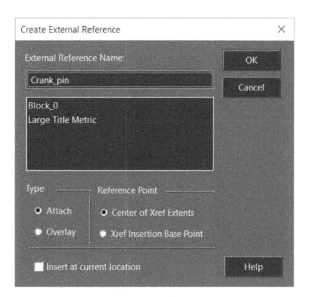

- Click **OK**; the crank pin will be displayed in the **Blocks** palette.
- Click and drag the Crank_pin from the Blocks palette and then release it into the graphics window.

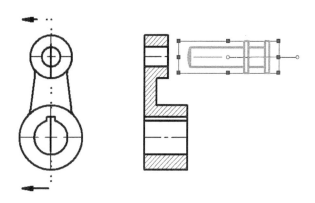

- Select Crank_pin.
- On the **Modify** toolbar, click **Transform** drop-down > **Move**.
- Select the intersection point between the centerline and the vertical line, as shown.

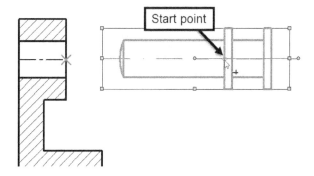

- Select the destination point on the crank, as shown.

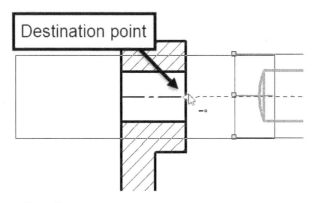

- Press Esc.

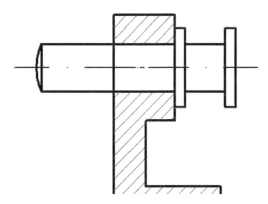

- Click **Insert > Create External Reference** on the menu bar; the **External Reference Files** dialog appears.
- Browse to the location of the **Nut.tcw** and double-click on it.
- Leave the default options on the **Create External Reference** dialog and click **OK**.
- Click and drag the **Nut.tcw** file from the Blocks palette into the graphics window.
- Rotate the Nut using the rotate handle, as shown.

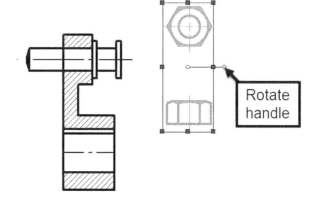

- Move the Nut, as shown.

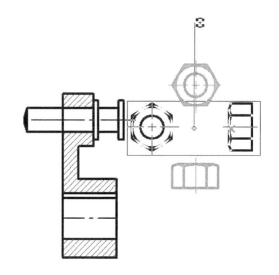

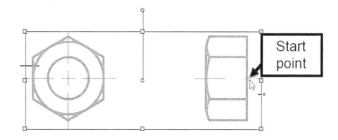

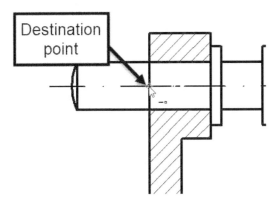

Clipping External References

You can hide the unwanted portion of an external reference by using the **XClip** tool.

- Select the **Nut** and click the **XClip** icon on the **Modify** toolbar.

- Select the **Rectangular** option from the Inspector Bar.

- Draw a rectangle as shown below; only the front view of the nut is visible, and the top view is hidden. Also, the clipping frame is visible.

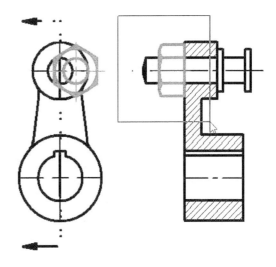

- Attach another instance of the **Nut**.
- Use the **Rotate** and **Move** tools to position the top view, as shown below.

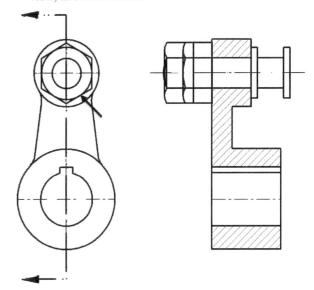

- Use the **XClip** tool and clip the Xref.

TurboCAD 2019 For Beginners

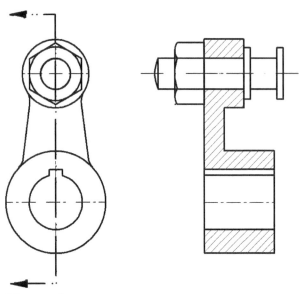

Exercise

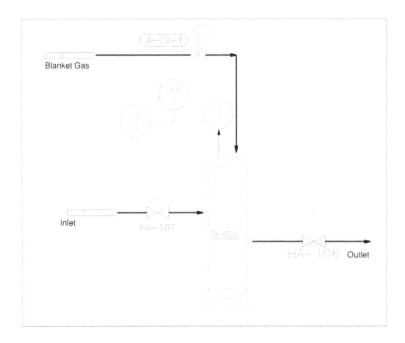

Chapter 9: Layouts & Printing

In this chapter, you will learn to do the following:

- **Create Layouts**
- **Specify the Layout settings**
- **Create Viewports in Layout**
- **Change Layer properties in Viewports**
- **Create a Title Block on the Layout**
- **Create Templates**
- **Plot/Print the drawing**

Drawing Layouts

There are two workspaces in TurboCAD: The Model space and the Layout. In the Model space, you create 2D drawings and 3D models. You can even plot drawings from the model space. However, it is difficult to plot drawings at a scale or if a drawing consists of multiple views arranged at different scales. For this purpose, we use Layouts or Layout. In Layouts or Layout, you can work on notes and annotations and perform the plotting or publishing operations. In Layouts, you can arrange a single view or multiple views of a drawing or multiple drawings by using Viewports. These viewports display drawings at specific scales on Layouts. They are mainly rectangular in shape, but you can also create circular and polygonal viewports. In this chapter, you will learn about viewports and various annotative objects.

Working with Layouts

Layouts represent the conventional drawing sheet. They are created to plot a drawing on a paper or in electronic form. A drawing can have multiple Layouts to print in different sheet formats. By default, there is one Layout available: Layout 1 and Layout 2. You can also create new Layouts by right-clicking on the Layout. Next, select **Insert** from the local menu. In the following example, you will create two Layouts, one representing the ISO A1 (841 X 594) sheet and another representing the ISO A4 (210 X 297) sheet.

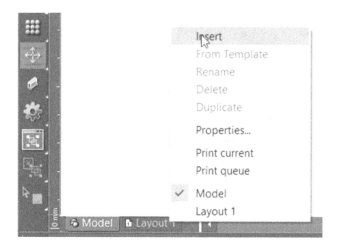

Example:

- Open a new drawing file using the Normal Metric template.
- Create layers with the following settings:

Layer	Linetype	Lineweight
Construction	Continuous	Default
Object Lines	Continuous	0.6mm
Hidden Lines	Hidden	0.3 mm
Center Lines	CENTER	Default
Dimensions	Continuous	Default
Title Block	Continuous	1.2mm
Viewport	Continuous	Default

- Create the drawing, as shown next. Do not add dimensions.

TurboCAD 2019 For Beginners

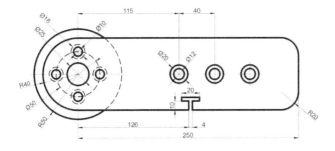

- Click the **Layout 1** tab at the bottom of the graphics window.

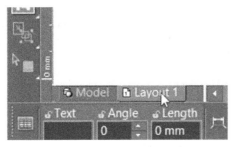

You will notice that a white paper is displayed with the viewport created automatically. The components of a Layout are shown in the figure below.

- Click **Insert > Layout > Properties** on the menu bar; the **Workspace Setup: Layout 1** dialog appears.
- Select the **Page Setup** option from the tree.
- Select a PDF printer from the **Printer/Plotter** drop-down.
- Select the **ISO A1 Size (594.00 x 841.00 mm)** from the **Workspace sheet properties** drop-down.
- Select the **Landscape** option.
- Select the **A1** option from the **Printer Paper** option.

- Set the **Plot Style table** to **acad.stb**.

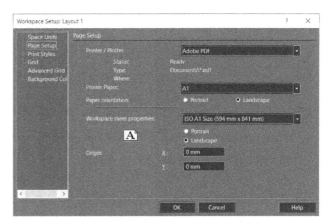

- Select the **Print Styles** option from the tree.
- Select the **Use named print style** option.

- Click **OK** on the **Workspace Setup: Layout 1** dialog.
- Right-click on the **Layout1** tab and enter **Rename**.
- Type **ISO A1** in the **Rename PaperSpace** dialog. Next click **OK**; the **Layout1** is renamed.

Creating Viewports in the Layout

The viewports that exist in the Layout are called floating viewports. This is because you can position them anywhere in the Layout and modify their shape size with respect to the Layout.

Creating a Viewport in the ISO A4 Layout

- Open the **ISO A1** Layout, if not already open.
- Click **View > Viewports > Viewport** on the menu bar.

Layouts & Printing

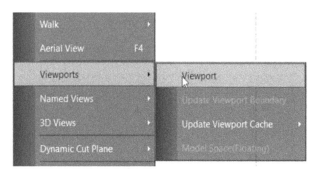

- Create the rectangular viewport by picking the first and second corner points, as shown in the figure.

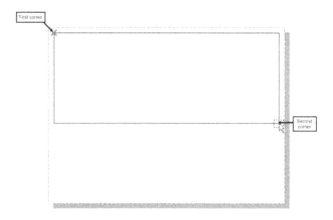

- Select **View_0** from the **Named View** dialog, and then click **Go To**.
- Click **Close** on the **Named View** dialog.
- On the Menu bar, click **Tools > Select**.
- Select the newly created viewport.

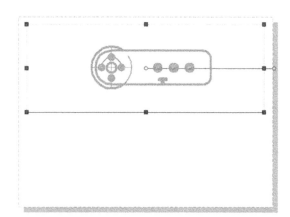

- Click **View > Viewports > Model Space (Floating)** on the menu bar; the model space inside the viewport will be activated. Also, the viewport frame will become thicker when you are in model space.

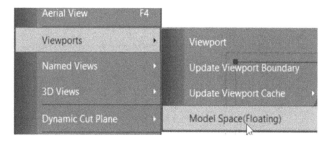

- On the menu bar, click **View > Zoom > Zoom Window**.

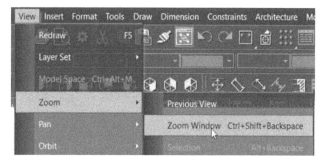

- Create a zoom window, as shown.

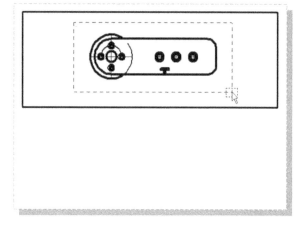

- Double-click outside the viewport to switch back to Layout.
- Use the **Circle** tool and create a 180 mm diameter circle on the Layout, as shown below.

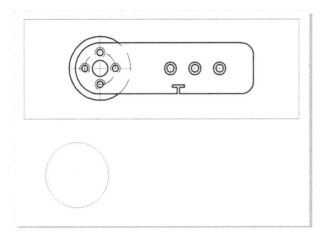

- Click **View > Viewports > Viewport** on the menu bar.
- Click the **Shaped Viewport** icon on the Inspection Bar.

- Select the circle from the Layout; it will be converted into a viewport.
- Select **View_0** from the **Named View** dialog, and then click **Go To**.
- Click **Close** on the **Named View** dialog.
- On the Menu bar, click **Tools > Select**.
- Select the newly created viewport.

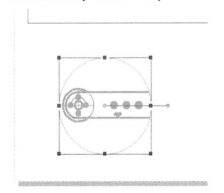

- Click **View > Viewports > Model Space (Floating)** on the menu bar
- Use the **Zoom In** tool to zoom into the drawing.

- Use the **Pan** tool and adjust the drawing, as shown below.

- Double-click outside the viewport to switch back to Layout.
- On the Menu bar, click **Tools > Select**.
- Select the rectangular viewport.
- Click the **Properties** icon on the **View** toolbar.

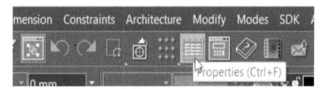

- Select the **Viewport** option from the tree located on the left side.
- Check the **Fixed** option in the **Scale** section.
- Type 2:1 in the **Scale** box available in the **Scale** section.

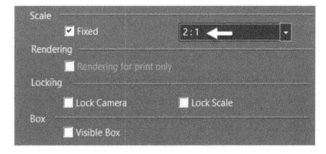

- Uncheck the **Viewport Box** option.
- Click **OK**.

Creating the Title Block on the Layout

You can draw objects on Layouts to create a title block, borders, and viewports. However, it is not recommended to draw the actual drawing on Layouts. You can also create dimensions on Layouts.

Example 1:

- Click the **ISO A1** Layout tab.
- Click the **Line** drop-down > **Rectangle** on the **Draw** toolbar.
- Set the **Title Block** layer as current.
- Select **By Layer** from the **Line Pattern** and **Line Width** drop-down available on the **Property** toolbar.
- Pick a point at the lower right corner of the Layout.
- Move the pointer toward the top-left corner.
- Press the TAB key and type **820** in the **Size A** box.
- Press the TAB key and type **580** in the **Size B** box.
- Press ENTER to create the rectangle.

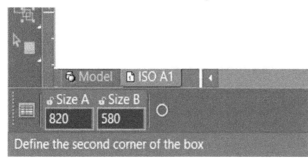

- On the menu bar, click **Insert** > **Blocks**; the **Block** palette appears.
- Select the Large Title block and click the **EC** icon on the Blocks palette.

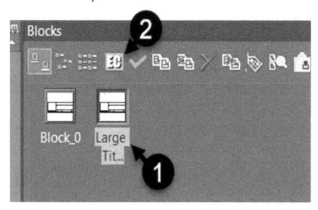

- Create attributes and place them inside the title, as shown below (refer to *Chapter 9: Blocks, Attributes, and External References* to learn how to create attributes).

- Click the **Finish Edit Content** icon on the **Blocks** palette.

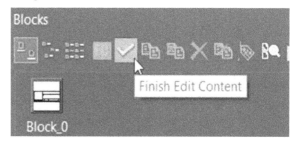

- Click and drag the Large Title block in the layout; the **Block Attributes Setting** dialog appears.
- Type **Layouts Example** in the **Attribute Value** box.
- Click **Next** and enter 1 as the DWG No. value.
- Click **Next** and enter 2:1 as the SCALE value.
- Click **Close**.
- Use the **Create Block** tool and convert it into a block.
- Select the Large Title block, if not already selected.
- Click the **Move selected entities** icon on the **Modify** toolbar.
- Deactivate the **Keep Original Object** icon on the Inspection Bar.
- Select the lower corner point of the large title block and the large rectangle.

TurboCAD 2019 For Beginners

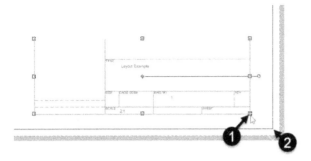

- Save the drawing file as **Layout-Example.dwg**.

Plotting/Printing the drawing

- Click the **ISO A1** Layout tab.
- Click the **Print Preview** button on the **Main** toolbar; the **Layouts Print Preview** window appears.

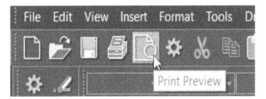

- Click the **Print Setup** button.
- Select the **A1** from the **Size** drop-down available in the **Paper** section.
- Click **OK**.
- Click the **Page Setup** button.
- Click the **Fit** button located at the bottom left corner; the preview window appears.
- Click **OK**, and then click **Print**; the drawing will be plotted.
- Click **Close** on the **Layouts Print Preview** window.
- Save and close the drawing file.

Exercises

Exercise 1

Create the drawing, as shown below. After creating the drawing, perform the following tasks:

- Create a Layout of the A3 size and then create a viewport.

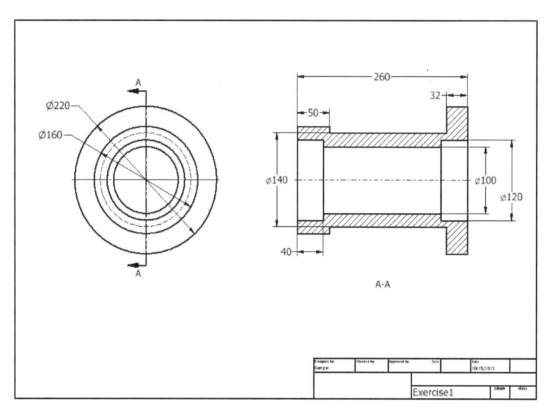

Exercise 2

Create and plot the drawing, as shown in the figure.

TurboCAD 2019 For Beginners

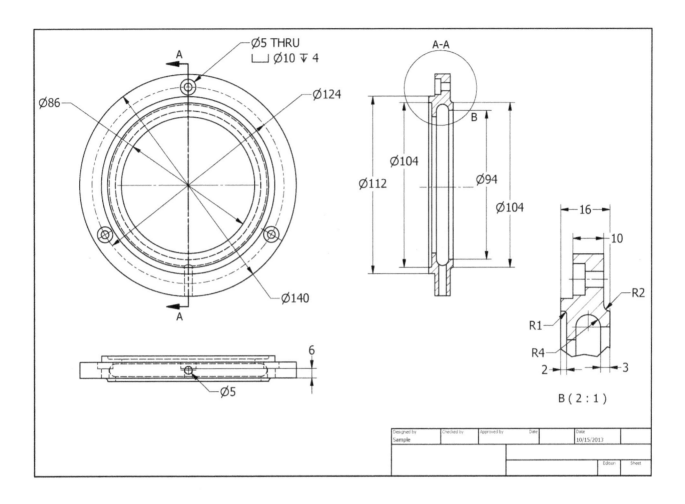

Layouts & Annotative Objects

Index

2-Point, 22
3-Point, 22
Angle, 37
Angular, 102, 109
Angular Construction Line, 86
Arc, 24
Arc Center and Radius, 24
Arc Tan to 2 entities, 53
Arc Tan to Line, 27
Array, 58
Attach, 142
Auxiliary Views, 81
Baseline, 100, 109
Block, 134
Block Attribute, 139
Block Editor, 138
Brush Pattern, 130
Center Line, 126
Center Mark, 103, 126
Center, Diameter, 21
Center, Radius, 55
Centerline, 104
Chamfer, 52
Circle, 21
Construction, 74
Continuous, 99
Create Block, 133
Create External Reference, 142, 143
Customize, 136
Datum, 105
Dialogs and Palettes, 7
Diameter, 102, 108
Double Point Arc, 26
EC, 138
Edit Control Points, 63
Edit Fit Point, 62
Edit Tool, 60
Ellipse, 30
Ellipse Fixed Ratio, 30
Elliptical Arc, 31
Explode, 54, 136
Extended Ortho, 128
Extents, 106
Fillet 2D, 51, 124
Finish Edit Content, 151
Format Painter, 117
Graphics Window, 4
Grid, 35
Help, 12
Horizontal Construction, 75
Horizontal Ray, 90
Inspector Bar, 6
Keep Original Object, 83
Keyboard Shortcuts, 12
Large Radius, 103
Layer Manager, 38
Line, 13
Load Folder, 137
Local Snaps, 39
Menu Bar, 4
Middle Point, 90
Mirror Copy, 52
Model Space (Floating), 149
Modify 2D Objects, 54
Move, 45
Named views, 91
New Layer, 38, 74
Object Trim, 56
Offset, 57
Offset Construction Line, 75

Ortho, 74
Orthogonal, 97
Page Setup, 152
Pan, 44
Parallel, 85, 98
Parallel Construction, 87
Pick Point Hatch, 121
Polygon, 18
Polyline, 20, 138
Previous View, 43
Print Preview, 152
Properties, 115, 148, 150
Quick, 105
Radial, 55
Radius, 102, 103
Rectangle, 16, 84
Rectangular, 150
Redo, 15
Replace References, 135
Rotate, 47, 83
Rotated Ellipse, 30, 90
Saving a drawing, 11
Scale, 48
Selection Modes, 9
Shaped Viewport, 150
Shortcut Menus, 7
Show Magnetic Point, 42
Shrink/Extend, 50

Smart, 96
Snap Modes, 41
Spline By Control Points, 29
Spline By Fit Points, 28
Starting TurboCAD, 1
Status bar, 6
Stretch, 54, 115
Style Manager Palette, 107
System requirements, 1
Tan Tan Radius, 49
Tolerance, 114
Tolerances, 110
Toolbar, 4
Trim, 49, 77
Undo, 15
user interface, 3
Vector Copy, 46
Vertical Construction Line, 74, 79
Vertical Ray, 89
Viewport, 148
XClip, 144
Zoom Extents, 43
Zoom Selection, 43
Zoom Window, 91, 149
Zoom-In, 43
Zoom-Out, 43
Zoom-Window, 43

CPSIA information can be obtained
at www.ICGtesting.com
Printed in the USA
BVHW011346301121
622868BV00008B/128